RETRO
WATCHES

Josh Sims and Mitch Greenblatt
Photographs by Tyler Little

RETRO WATCHES

The Modern Collector's Guide

Over 300 illustrations

CONTENTS

- 6 INTRODUCTION

THE WATCHES
- 10 Advance/Modulus Driver's Watch
- 12 Amida/Hudson Digitrend Jump Hour
- 14 Amyria Chronograph
- 16 Benrus Citation Electronic
- 18 Benrus Pop-up LED Sideview
- 20 Buler Century 2000 and Buler 'Green Machine'
- 24 Bulova Accutron 'Woody'
- 26 Bulova 'Big Block' Accuquartz
- 28 Bulova Christian Dior Collection
- 30 Bulova Computron Driver's Watch
- 32 Candino Masternove Automatic
- 34 Caravelle Bullion
- 36 Clipper Alldate
- 38 Codhor Chronograph
- 40 Damas Mechanical Jump Hour Cuff and Damas Mechanical Jump Hour Vertical
- 44 Derby Swissonic
- 46 Desotos 'Cuffbuster' Chronograph
- 48 Dugena Chronograph
- 50 Space Age Design
- 54 Dynasty Mido Swissonic
- 56 Elgin 'Golf Ball'
- 58 Enicar 340 and Enicar Sherpa 320
- 62 Eterna Driver's LED
- 64 Favre-Leuba Moon Raider
- 66 Girard-Perregaux Casquette
- 68 Gubelin G-Quartz Cuff
- 70 Hamilton Dateline TM-5903
- 72 Hamilton Electric Victor
- 74 Hamilton Electronic
- 76 Hamilton Fontainebleau B and Hamilton Fontainebleau Chrono-Matic
- 80 Hamilton Odyssee 2001
- 84 Hamilton QED LED
- 86 Hamilton Self-winding
- 88 Hamilton Thin-o-Matic T-403 'Shark Fin'
- 90 Handcuff Bracelet Watch
- 92 Heuer Ford RS Split Lap Unit 77 LCD Chronograph
- 94 Hudson Directime and Itraco Directime

98 The Retro Aesthetic	148 Omax Driver's Watch	208 Tell Fleurier Swissonic
102 Jaeger-LeCoultre 'Disco Volante'	150 Omega Driver's Watch	210 Teviot Jump Hour Mechanical Digital
104 Jaeger-LeCoultre Master-Quartz Models	152 Pallas Quartz and Quartz Segtronic	212 Timshel Mechanical Digital
108 Jovial Vision 2000	156 Paul Smith Driver's Sideview	214 Tissot Synthetic Idea 2001
110 Jules Jürgensen Wedge	158 Pierre Balmain	216 Titus Domed Automatic
112 LeCoultre HPG Memovox	164 Pierre Cardin Espace Collection	218 Design Details
114 Lip Baschmakoff Jump Hour	172 Pirofa 'Bullhead' Chronograph	222 Tosca Driver's Watches
116 Lip by Isabelle Hebey and Miscellaneous Mechanical Lip Models	174 Mechanical Vs Quartz	226 Union Ladies' Watch
	178 Prisma GT Valjoux Chronograph	228 Universal Geneve Tank
	180 Racing Helmet Vertical Drum Display	230 Vulcain 'Eye' Dial
124 Lip Mach 2000 Chronograph and Lip Mach 2000	182 Rado NCC 404	232 Vulcain Jump Hour
	184 Record (Longines) Plastic Automatic	234 Waltham Jump Hour Chronograph
128 Lip Secteur		236 Webster Hand-Wound
132 Lip Skipper	186 Rolex Midas Cellini	238 Wittnauer Futurama
134 Lord Nelson Mystery and Lord Nelson Jump Hour	188 Rolex Oyster	240 Wittnauer Polara
	190 Royce 'Mexico' World Cup Watch	242 Yema Mechanical Digital
136 Lord Nelson Trapezoid	192 Sheffield Outsized and Sheffield Jump Hour	246 Zenith Time Command Futur
138 Lucerne 'D' Jump Hour		248 Zodiac Astro
140 Mido Cushion Automatic	196 Sicura Jump Hour	
142 Mortima Mayerling Thermo-Compass 'Survivor' Watch	198 Sorna Mechanical Digital	250 **ABOUT THE COLLECTOR**
	200 Spaceman Watches	251 **ACKNOWLEDGMENTS**
	204 Sultana Double Retrograde	252 **A CHRONOLOGY OF WATCHMAKING**
144 Movado Zenith Wood Dial	206 Swank Watch	
146 Nobreza Mechanical Digital		255 **ILLUSTRATION CREDITS**

INTRODUCTION

The watch world is quickly caught up in a fascination with movement – the inner workings of a mechanical watch – often at the expense of the exterior. But it is the aesthetic of a watch that the wearer has a lasting relationship with, not its precision, nor even really its other, typically redundant, functions, however sophisticated all those little cogs and wheels within may be. It is the form, the colour, the texture, the finish and the details that speak to many: watch nerds of another kind, unbowed by haute horlogerie, but blown away by a striking face.

Arguably, these design qualities spoke loudest during the 1960s and 1970s, watchmaking's golden age of experimentation in the avant-garde. This period may not have produced out-and-out watch classics – designs acclaimed as timeless, historic, mimicked by copyists ever since. Indeed, many of their makers themselves were short-lived. But it did produce watches that were, put simply, very cool.

Cool, indeed, in a 'straight from the fridge, daddio' way. Cool also because they are unlike almost anything available new on the market today: oddities bearing unfamiliar names, from a narrow window of unfamiliar times – the late 1960s and early 1970s – made in low numbers, in a left-field style that had fallen out of favour by the end of that decade, leaving unsold inventory ready to be snapped up by collectors hungry for watches with a difference.

From the collection of watch aficionado Mitch Greenblatt – from which comes the selection of watches in this book – take, for example, a 1970 Sultana, with its double retrograde display, 'which makes it hard to tell the time,' he concedes; the prism that corrects the time display on the fantastically named Amida/Hudson Digitrend LRD Light Reflective Display from 1974, 'emulating digital watches in a totally different way'; or the 1972 Desotos 'Cuffbuster' piece, 'which you have to see to really believe,' he says. 'It's a ridiculous, cumbersome design – a giant Cadillac on your wrist. It's just the strangest. And for some people the ugliest...' But certainly striking, fresh and a product of its time.

Assay a summary of the characteristics of the more striking watches of the turn of the 1970s and one might stress their asymmetric case shapes; their deep, outsized case sizes; the bright, bold colours of their dials and numerals; their use of brushed steel and chrome, rarely gold; their hunky bracelets; and, perhaps above all, their often innovative time display methods, with direct and jumping hours especially typical – ease of reading the time often being secondary to the visual effect.

Of course, such bold designs had been attempted on rare occasion before: Elgin's Lord Elgin of 1950, with its all-metal face and tiny time window, for instance, or Kienzle's boxy Life watch of the late 1950s. Most strikingly perhaps, in 1957, Hamilton set new style benchmarks with the world's first electric watch, its Ventura, a futuristic, triangular model designed by Richard Arbib and still in production (outselling modern models, in fact, with a replica launching next year). Hamilton followed it with the equally bold Spectra (1957), Saturn (1960) and Savitar (1961) models, winning the company the contract to design a watch more literally of the future, for the cast of 1968's *2001: A Space Odyssey*, a watch that was originally planned to go to market after the film's release. Indeed, the movie, as with the watches of those years, was also a reflection of arguably one of the culturally most far-reaching (literally and figuratively) eras of the twentieth century: the space race.

Inspired by the leap for the moon, suddenly television sets looked like astronauts' helmets, record players like robots, chairs like domestic planetoids. Bright, curvaceous, shiny and organic became the watchwords of a retro-futuristic design ethos, pioneered by the likes of Joe Colombo, Vico Magistretti and Eero Aarnio, and emphatically new. Watches, similarly, no longer looked to the past. They, like other space-inspired designs of the time, offered hope through materials and experimentation. 'Original space programme design was fat-free, but had glamour embedded in it – there is glamour in technology. In fact, what the space age aesthetic really

OPPOSITE Arguably the best, and most timeless, of LED watch designs to date – the stainless-steel version of Girard-Perregaux's Casquette (1976).

ABOVE Zodiac's Astro of 1969 was a striking example of a so-called mystery dial, the hands 'floating' on transparent concentric discs.

did was break free of aesthetics,' as the product designer Ross Lovegrove, who has designed for Tag Heuer, has it.

The period was a golden age both for its emphasis on progressive case and dial design over movement and for the way it saw watch design become part of (rather than operate separately from) the broader design ethos of the times. Indeed, some of today's major watch names offered many of the more distinctive pieces of the period: models from the then new company Seiko and the ever-collectible Lip, Breitling's Chronomatic, Omega's Chronoquartz, Girard-Perregaux's Casquette LED, Jaeger-LeCoultre's Memovox, Bulova's Spaceview, Longines' Wittnauer Futurama or Comet Mysterieuse, Tag Heuer's Monaco...' The social changes of the time were simply reaching a peak in their readiness to accept the more daring, provocative, aggressive, and watch design reflected that,' says Matthias Breschan, president of Hamilton.

But it was more typically the small, often family-owned and now mostly defunct watch companies – the likes of Lanco, Microsonic, Iza, Le Phare, Itraco and Gigandet, among many others – that began to look to industrial and product designers for their watch ideas rather than rely on in-house specialists. Roger Tallon, for example, is as well known for designing portable televisions and the first French TGV trains as for the 1970s' cream of Lip. Companies better known for more utilitarian engineering, such as Braun and even Hughes Aircraft, launched watches. The period also saw the advent of the fashion designer watch: Pierre Cardin and Pierre Balmain, both determined futurists, were among the first to license their names to watch designs. Even artists got in on the act: in 1971 Tian Harlan developed his 'chromachron' dial, displaying time by way of variously coloured segments. It was three oranges past pink.

Of course, 'retro' – at least when done well – has become a strong force in design, and that builds interest in such watches. But these designs were really pioneering, resulting in styles that, undoubtedly, are too strong for some people. It is certainly a niche market – these designs are a long way from the classics of watch design typically cited, from Omega's Speedmaster to Audemars Piguet's Royal Oak, Cartier's Tank to Rolex's Submariner – but, undeniably, an influential one. Whenever a modern company launches a genuinely distinctive watch it tends to take the 1970s as its reference point.

They brought with them some genuine innovation: sometimes in mechanics – the Jaz Derby Swissonic, for example, displayed the time using cylindrical-barrelled rolling wheels powered by a transistorized system – but always in looks. Cardin's Jaeger-signed crescent cross-section-shaped

The 1960s and 1970s saw many experiments with the display format on watches, with this Lip Secteur retrograde watch seemingly taking its inspiration from automotive fuel gauges.

watch of 1971 is angled on the wrist to make reading the time easier while driving; that Desotos was shaped to make reading the time possible without pulling back your shirt cuff.

Just as importantly, loud was in demand. Sixties and early 1970s modernism, be it in furniture, fashion or automotive design, was in part a counteraction to drab post- World War II privations and, in Europe, reflected a pursuit of the stylistic boom years being enjoyed in America, where household consumer goods were going hi-tech, and neon, neo-Art Deco candy colours and the decidedly outsized suggested an altogether more upbeat, larger life. Compared with the seemingly ceaseless allure of the 1960s' youth-driven avant-garde, the 1970s may be commonly condemned as the decade that taste forgot, but it was also one of experimental extremes: flares and lapels were not the only things that were big.

These were strong watch designs for which innovation and style went hand in glove. They neither placed too much emphasis on the technical nor were just flat circular objects with Roman numerals and hands that went round, the basic characteristics of most classic watches. Their outlandishness was part of their appeal. And that remains the case: there is a demand for distinctiveness, but it is hard to find outside of the world of high-end watchmaking today.

Sadly, with the fizzling out of the space race in 1972 – that December saw the last manned mission to the moon – so, with a whimper, did demand for such show-stopping 'retro' watches fade too. Just as everything came together at the right moment for this special time in watch design – particularly the birth of consumer culture, the first time people started owning more than one watch, the first time the watch came to be regarded as more than functional – so the times also brought it to an end. The economic crisis of the 1970s, and the watch industry crisis of quartz, together saw watch design place a renewed emphasis on even greater levels of functionality. It was all about watches with calculators, altimeters, gizmos over aesthetic impact. Or – a word not often used in high-end watch design – fun.

In many respects watch design of the period was exemplary of a brief stylistic adventurousness that has since been lost to all but the rarefied world of concept watchmaking, but it says something that the Swiss watch industry has, over recent years, seen a challenge to its innate conservatism in the shape of a still fledgling but growing sector of independent watchmakers pursuing the aesthetic and engineering extremes. Count among these Max Busser's Horological Machine 2, Cabestan's Winch Tourbillon, the Urwerk 202, with its display on three orbiting satellites, and Alexandros Stasinopoulos's asterisk-shaped Ora, with its tape measure-style display, as well as pieces from the likes of HYT and Richard Mille, among others. There are, occasionally, watches designed outside of normal watch circles, for example by Marc Newson, Ettore Sottsass, Paul Smith or TokyoFlash, offering a more dynamic styling too.

Indeed, it might well be argued that, with a renewed demand for individual expression, together with designers who grew up during the 1960s and 1970s coming into positions of influence and free to give vent to their nostalgia for the space aesthetic or *Star Wars*, we may yet see watches boldly going where none has gone before. For the time being, however, constrained by commercialism, cornered by brand expectations or restrained by a classicist vision of what a prestige timepiece should look like, the mainstream arguably lacks a visual appeal that, as cellphones steadily replace watches as time-tellers, is all the more required.

It might be concluded that too many of the big brands have a corporate mindset that leaves little room for the innovation of the 1970s. There is a reason that these 'retro' watches now sell to the likes of artists, pop stars, designers and fashion people: they are creative and are looking for something that expresses that sensibility. And that is what the watches of the space race era could do.

Josh Sims

ADVANCE / MODULUS DRIVER'S WATCH

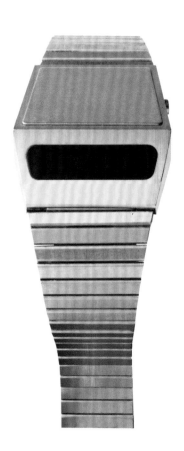

If gold has typically been associated with classicism in watch design – round cases, Roman indices, leather straps and so on – then that is clearly not the case with this rare driver's watch. Marketed by two brands, Advance and Modulus, and with an American-made electronic movement, the model was short-lived, but stood out all the same. This was in part down to the LED display, temporarily displayed at the touch of a button, but also to the thin plate that sweeps away behind the dial. Remove this and it could be customized, as some owners did. Most examples are now found with some kind of pattern etched or stamped into this section of the case.

YEAR OF RELEASE 1974

MOVEMENT Solid state electronic light-emitting diode module set

RELATIVE VALUE ★ ★ ★

NOTABLE FEATURE Owners often customized the case

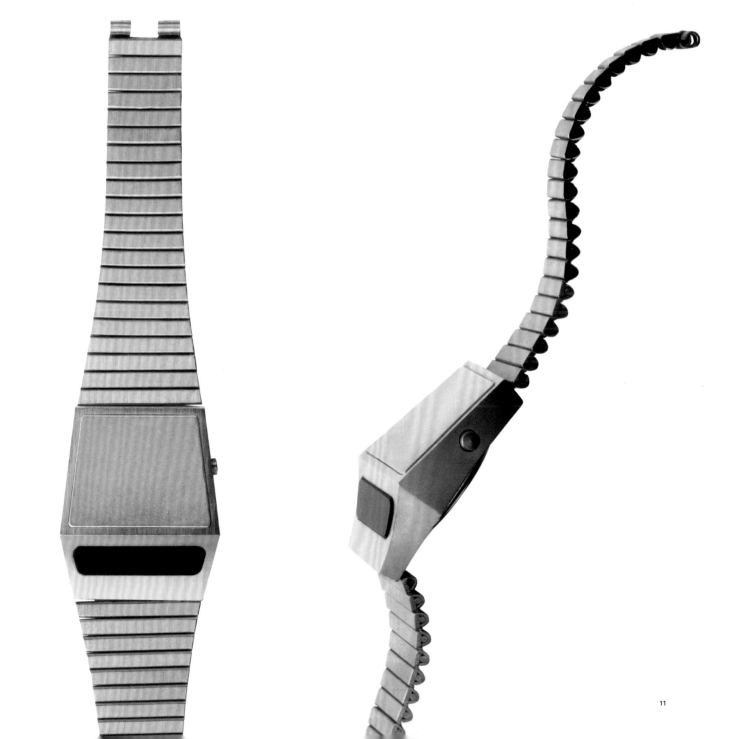

AMIDA / HUDSON DIGITREND JUMP HOUR

YEAR OF RELEASE 1974

MOVEMENT MBWC 420 calibre mechanical

RELATIVE VALUE ★ ★ ★

NOTABLE FEATURE Featured an unusual LRD, or light-reflecting display, system

Counterintuitive at first glance – on the wrist, the angles all seem wrong for effective viewing of the dial – the Amida/Hudson Digitrend was one of a new generation of 1970s driving watches, with the dial situated so as to allow the time to be read without taking your hand off the steering wheel. There were a number of LED watches with this feature produced throughout the decade, including the Mido Swissonic, the Bulova Computron and the Girard-Perregaux Casquette. There was, indeed, something of a digi-trend going on. And the Amida was, as the company put it in an ad for the Baselworld watch industry fair of April 1976, one of a 'new generation of digitals'. Although the patent for the watch was filed by Amida in 1970, it did not go into full production until after this Baselworld unveiling.

Certainly the Amida/Hudson Digitrend was doing things differently. The watch, designed by Joseph Barnat, offered a manual-wind mechanical alternative with what it called its LRD, or light-reflecting display, to compete with those LED (light-emitting diode) displays. The LRD system was more complex than it might at first seem. The numerals – printed in orange to create a stylistic passing nod to the vibrant, typically red hues of an LED display – were actually printed backwards on a rotating disc, and then reflected sideways through a clear plastic prism, which corrected them for the display. The jump hour was printed on a horizontal mirrored disc and then optically corrected so as to appear vertical in the display. So there is a kind of double illusion at play.

Despite the complexity of the innards, the chrome-plated base metal case (painted black in some models) and the manual-wind movement kept the price low, especially for some versions. While the one pictured has a rare 17-jewel movement, many of Amida/Hudson Digitrend watches were produced with a very basic 1-jewel movement made by the Michael Berger Watch Company, whose business encompassed the production of near-disposable watches for fashion brands of the time. Yet, regardless of the build quality, the Digitrend, with its blocky yet aerodynamic form and curious display, was one of the most futuristically styled watches of the decade.

AMYRIA CHRONOGRAPH

With sub-dials to squeeze side-by-side into a circular dial, a more typical chronograph watch is always a challenge for a watch designer; it is all too easy to end up with a dial that looks cluttered. One solution would be to use a square case, as with Heuer's iconic Monaco model, another an elliptical, 'disco volante' or flying-saucer-shaped case – both giving more space along the x-axis. Few watches followed this second path, yet this Amyria model, with a Valjoux 7734 movement, wisely did so, allowing for a pleasing symmetry between the upper and lower indices. Appropriately enough, the case shape somewhat gives the impression of a wheel passing at speed too.

YEAR OF RELEASE 1974

MOVEMENT Valjoux 7734

RELATIVE VALUE ★ ★ ★

NOTABLE FEATURE Uncommon ellipse casing

BENRUS CITATION ELECTRONIC

There was a time even before the advent of quartz when prestige Swiss watches were not all associated with mechanical movements; when, for a short period at least, some watches would proudly display on their dials, as in this case, the fact that they were 'electronic', with all that that suggested of the bright, bold future. Indeed, this Benrus Citation came with an ESA 9154 Dynotron electro-mechanical movement, a hybrid of battery, tuning fork or balance wheel with a transistor of electrical contacts of the kind that, along with electric movements, was popular only between the late 1950s and early 1970s. It was during this time that transistors were available at the scale and price required for their use in watches, but when the fledgling quartz movement had yet to go mass-market.

Other watch designs may have pioneered this brief moment in horological history – the Bulova Accutron was the first watch to use a transistor and the ESA Dynotron the first to use a balance wheel with a transistor, while Hamilton can lay claim to the first strictly electric watch. But this minimalistic model from Benrus is no less striking for that. Benrus, established in 1921 and named after its founder, Benjamin Lazarus, was the third biggest in the USA after Bulova and Hamilton; it was the advent of quartz movements that put paid to the company, which filed for bankruptcy in 1977, just two years after this watch's release.

YEAR OF RELEASE 1975

MOVEMENT ESA 9154 Dynotron electro-mechanical

RELATIVE VALUE ★ ★ ★

NOTABLE FEATURE Benrus was once the USA's third-largest watchmaker

BENRUS POP-UP
LED SIDEVIEW

Throughout the 1970s, LED display watches suggested that the wristwatch was on the brink of becoming something closer to the wrist computer. While watches with in-built calculators followed from such companies as Hewlett Packard, Uranus Electronic and Pulsar ('surprisingly useful to the top executive', as one Pulsar print advertisement had it), computing power did not – and would not for decades, until the advent of WiFi connectivity and the smartwatch.

But, stylistically, the use of LED seemed to encourage a new adventurousness, in part perhaps because the time display itself took up so little of the overall case area. That

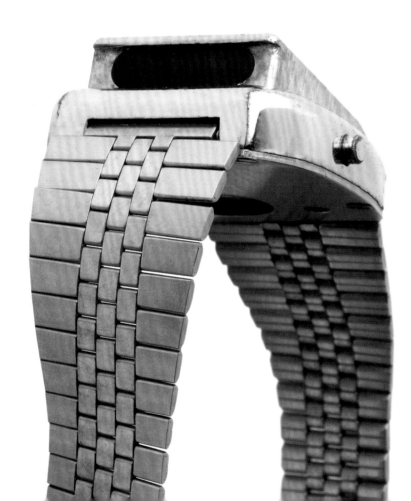

meant the rest could be used to new functional ends, as in the panels that drive the Roger Riehl-designed Synchronar, the first solar-powered watch, or to more expressive ends. Omega's rectangular-cased TC-1 is a case in point, as are the one-off designs by 1970s jeweller Andrew Grima, with their stepped cases or cases mimicking belt buckles. But just as striking, for its absolute minimalism, is Benrus's pop-up sideview model. Closed, the watch is plain and bracelet-like. Only when a button is pressed does the time display reveal itself.

YEAR OF RELEASE 1971

MOVEMENT Techniquartz module

RELATIVE VALUE ★ ★ ★

NOTABLE FEATURE This early LED model hides the time display until needed

BULER CENTURY 2000
+ BULER 'GREEN MACHINE'

The name of the Century 2000 suggests a time some three decades from its date of issue, reflecting the aspiration Buler had in designing the model as it did – with an automatic movement powering a jumping hour display with hours, minutes and seconds set on mesmerically rotating discs, all set in an unusual concave stainless-steel case. Pushing against the conservatism inherent in many historic Swiss brands had been Buler's raison d'être when it was founded in 1945 by Charles and Albert Buhler.

It worked for a while too: demand was such that it had to twice expand its manufacturing plant over the 1950s and 1960s. Such watches as the unusual 'Green Machine' followed, an outsized Art Deco-inspired piece with decorated case, hexagonal crown and a green leather and woven fabric strap. As with so many smaller, mechanical-movement watchmakers, sales petered out with the advent of quartz movements, until the company was finally taken over by a private entrepreneur in 1990 and underwent a relaunch. Arguably, however, the Century 2000 remained the brand's stand-out design.

YEAR OF RELEASE 1971

MOVEMENT Swiss manual-winding mechanism

RELATIVE VALUE ★ ★ ★

NOTABLE FEATURE Buler made its name by pushing against typical Swiss conservatism

22　BULER CENTURY 2000 + BULER 'GREEN MACHINE'

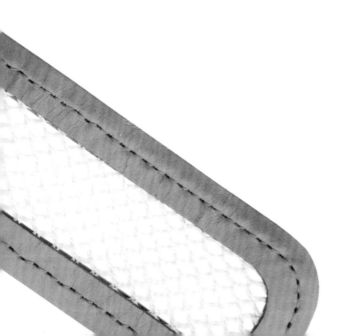

YEAR OF RELEASE 1971

MOVEMENT Swiss manual-winding mechanism

RELATIVE VALUE ★ ★ ★

NOTABLE FEATURE Relatively unknown and rare fashion design from Buler

BULOVA ACCUTRON 'WOODY'

The use of wood in watches, primarily for the case and dial, was, by the early years of the twenty-first century, something of a trend. Riding the wave of a growing eco-consciousness perhaps, wooden parts in what had to date been most familiar in cold hard steel had a certain green appeal in pieces from such diverse makers as Urwerk and Bell & Ross, as well as numerous small brands (quite aside from the *métier d'art* use of wood in marquetry dials by Cartier and Patek Philippe, among others).

But this was not a new idea. Tapping a 1950s/60s mid-century design preference for natural materials in interiors – leather, bent plywood, oak – Hamilton, for example, prototyped its Flight II and Pacer Electric models with wooden dials, finally producing a number in the guise of its (aptly named) Sherwood collection, with yellow gold cases and Mexican mahogany dials. Likewise many other brands, from Universal Genève to LIP, Rolex to Movado/Zenith, experimented with wood throughout the 1970s. Some would use wood just for an accent, as with the wooden bezel and bracelet insets in this Bulova Accutron model (with Bulova 2182 movement). It was nicknamed the 'Woody' – a moniker that would come to be typically applied to any model unusual enough to use wood.

YEAR OF RELEASE 1973

MOVEMENT Accutron tuning fork, marked Bulova 2182, USA

RELATIVE VALUE ★ ★ ★

NOTABLE FEATURE Use of wooden insets for bracelet and bezel

BULOVA 'BIG BLOCK' ACCUQUARTZ

Outsized proportions have always been one of the dynamics watch designers like to play with. The so-called 'Big Block' (or calibre 228) version of the Bulova Accutron was one of the first watches to do so, with its unforgivingly stark, massive – not to mention hefty – solid steel case so large that its manufacture had to be outsourced to Howard Hughes's Hughes Aircraft. Hughes Electronics also produced the movement and LED display, which has an unusual dot matrix arrangement for the numbers; later versions used the more commonplace 'bar' format.

This model was sold for some US$250 new, suggesting that Bulova was pitching it as a premium product. Within a few years LED watches would be available from such companies as Texas Instruments for a tenth of that price, or even cheaper in plastic cases. LED displays would prove short-lived anyway, soon made outmoded by LCDs (liquid-crystal displays), which gave constant readouts and at the same time a longer battery life. All the same, for obvious reasons, this 'Big Block' makes a lasting impression.

YEAR OF RELEASE 1974

MOVEMENT Hughes Aircraft light-emitting diode module (renamed Bulova calibre 228)

RELATIVE VALUE ★ ★ ★

NOTABLE FEATURE Use of dot matrix as opposed to bar format for the LED display

BULOVA CHRISTIAN DIOR COLLECTION

'Inside every Dior body, there beats a heart of solid Bulova,' ran the odd copy for an advertisement in 1972. 'Christian Dior watches are tailored to fit your wrist,' offered a second ad, announcing the notion that, when you buy one of said watches, the jeweller will alter the band to better fit you, or return it to Bulova if necessary to get it just so. 'Yes, it's an uncommon amount of attention to devote to fitting a watch. But then, these are very uncommon watches,' the ad notes.

These manual-winding watches pictured – with a more convenient leather strap – were just two of some fifty designs commissioned by Bulova from the French fashion house, including cases hexagonal, rectangular and boldly asymmetric, with bracelet straps (mostly in rolled gold plate) woven and roped, among many other styles. This was one of the first collaborations between a watchmaker and a fashion brand, a mutually beneficial marketing plan that would subsequently become a commonplace part of the consumer landscape.

YEAR OF RELEASE 1972

MOVEMENT Swiss manual-winding mechanical

RELATIVE VALUE ★ ★ ★

NOTABLE FEATURE Just one of 50-plus designs commissioned by Bulova from Dior

BULOVA COMPUTRON DRIVER'S WATCH

When the Bulova Computron was released it was considered decidedly unconventional, in large part because of its combination of a trapezoidal case, cross-hatched finishing and a driver's sideview display (Bulova was so enamoured of the design that in 2019 it would reissue it with a ridged finish). The logic of the design may have been somewhat flawed: the display is angled to prevent the reflected glare of direct sunlight dazzling the car driver, and so that they do not have to twist their wrist away from the steering wheel to see it, but then they have to take one hand off to press the button that activates the display anyway– a display in a case with a highly reflective gold metallic surface.

This from a nation that had, just seven years previously, put a man on the moon, and from a watchmaker with an illustrious history dating to 1875. Add in that LED was, as a technology, already on its way out at the time the watch was released, and the Computron might be considered something of a white elephant. And yet, while not a practical design success, its undoubted aesthetic distinctiveness made the original Computron one of the most revered of all LED watches.

YEAR OF RELEASE 1975-76

MOVEMENT Quartz module

RELATIVE VALUE ★ ★ ★

NOTABLE FEATURE One of the last LED production watches, before LCD came in

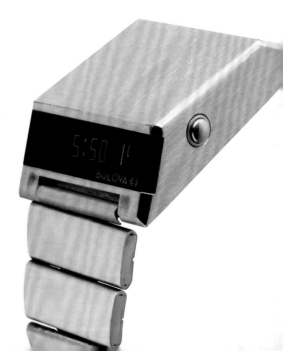

CANDINO MASTERNOVE AUTOMATIC

Candino was founded in Herbetswil, Switzerland, by Adolf Flury-Hug in 1947, a family firm building on the postwar reconstruction taking place across Europe. Success did not come quickly; it would take almost a decade before the company constructed its own factory, and it would be 1971 before it started production of automatic watches. Candino did, however, read the zeitgeist well. Rather than pursue the market for mechanical watches – one already looking to be under threat thanks to the advent of the seemingly much more progressive quartz movement – by 1976 it had come to be regarded as a specialist in electronic digital watches, with designs akin to a number of Casio classics that followed. Four years later it brought in what it called an 'ultramodern' assembly line for the production of extra-slim models.

Unfortunately – at least as far as the interest of this book goes – by the early 1980s Candino had retained its use of quartz movements but moved towards a much more classic styling, including the use of Roman numeral indices and the production of such complications as moon phase indicators. It did not leave all technical advances behind it: in 1987 it produced the first watches to feature a compass (the top half of the case slides aside to reveal the compass below), and the following year introduced the first two-tone ceramic watch, using

YEAR OF RELEASE 1973

MOVEMENT Swiss automatic

RELATIVE VALUE ★ ★ ★

NOTABLE FEATURE Candino also produced the first compass watch

a material called Innotec. The year after that it brought out a watch with the world's smallest mechanical altimeter.

Stylistically, however, Candino's most distinctive pieces belonged to the 1970s, including this extremely rare Masternove watch, with only a little over half of its almost round case actually used for time display and, indeed, reducing the indices for four to eight o'clock to a single line along the bottom of the starburst fumé dial. This truncation gives the Masternove the feeling of a retrograde watch. The bottom third of the case does not go entirely to waste – there is a small date window set dramatically at its centre. The second hand can briefly be seen to sweep under the steel covering via this same window.

CARAVELLE BULLION

Playfulness in the visual language of a watch design is a rare thing in itself, the Swiss business in particular tending to be overly serious and too conscious of the need to give its expensive products a longevity that humour perhaps cuts against. But look at the chunky case on this green-dialled, Corfam-strap, Art Deco-inspired watch from Caravelle, an American maker launched in 1962. Those four sections are surely meant to remind the viewer of gold bars lined up alongside each other. What else then to call this manual-wind model but the Bullion?

YEAR OF RELEASE 1970

MOVEMENT Swiss manual-winding mechanical

RELATIVE VALUE ★ ★ ★

NOTABLE FEATURE The watch takes its name from its distinctive shape

CLIPPER ALLDATE

YEAR OF RELEASE 1970s

MOVEMENT German Unisonic HB 313 automatic

RELATIVE VALUE ★ ★ ★

NOTABLE FEATURE Unusual day/date display bezel

It is arguably the most basic of complications – as watchmakers refer to a watch's functions in addition to telling the time – but that is not to say that watchmakers have not sought new ways of displaying the date: in a window, under a magnifying glass bubble, via a pointing hand, and so on. The Clipper Alldate (with a German Unisonic HB 313 automatic movement) may be small, at just 38 mm (1½ in.) across, but it features one of the more interesting – if questionably useful – functions of watches of the 1970s. Turning the bezel allowed the display of the day for each date of the current month. The use of colour in this design is unusual too, with two near-quarters of the dial in a purely decorative rich marine blue, matched to the strap in Corfam, a synthetic leather substitute.

CODHOR CHRONOGRAPH

Few watches are more emblematic of the 1960s' and 1970s' emphasis on strong aesthetics in watch design than this chronograph, with its manual-winding Valjoux 7734 movement, from manufacturer Montres Codhor of Sarcelles, France. Two atypical shapes sit not in harmony but in combat – the tonneau case and the diamond-shaped dial. But, visually, the clash works.

YEAR OF RELEASE 1970s

MOVEMENT Manual-winding Valjoux 7734 mechanical

RELATIVE VALUE ★ ★ ★

NOTABLE FEATURE A clash of tonneau and diamond shapes

DAMAS MECHANICAL JUMP HOUR CUFF
+ DAMAS MECHANICAL JUMP HOUR VERTICAL

Another of the many Swiss makers of mechanical watches to succumb to the advent of cheaper, more accurate quartz movements, Beguelin & Cie (established 1903) nevertheless created a number of visually arresting pieces under its Damas brand, active from the 1950s and created specifically for the Middle East market. (Damas would survive as a brand fronting a jewelry retail chain in the region.) The cuff designs of both of these pieces – one with a jumping hour, the other with the mechanical digital display more unusually set vertically – have a space age quality to them, in part due to the expanses of silvery brushed stainless steel.

YEAR OF RELEASE 1970s

MOVEMENT Swiss automatic winding

RELATIVE VALUE ★ ★ ★

NOTABLE FEATURE The Damas brand was created for the Middle East market

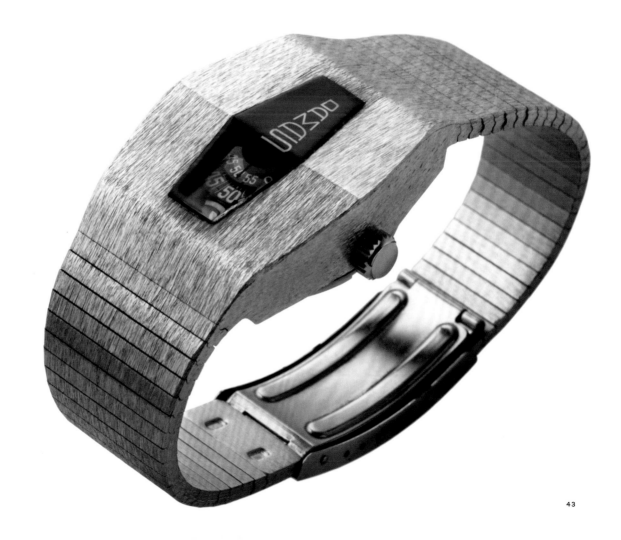

DERBY SWISSONIC

From the outside the Swissonic, a driver's watch from Derby, all is serene. But the clarity of the black and white dial and all that brushed-steel casing hides a world of complexity below. Indeed, it has been described as housing the most awkward jump minute movement – not jump hour, but jump minute – devised for mass production. And one that uses fragile plastic parts at that – all very forward-thinking at the time, but in retrospect making for a bold experiment rather than a successful watch, at least in terms of operating life.

In brief, inside is a speedometer drum powered by an ESA transistorized electromagnetic module. The battery powers the electromagnetic coil that operates the balance wheel via magnets, while the roller on the balance moves an escapement wheel that transfers torque to a metal rod on the drum, which is packed with gears. These move the seconds drum, which rotates the minute drum every sixty seconds, a second minute drum every ten minutes and all of the drums every sixty minutes. There is a minuscule spring within the seconds drum that makes all this possible – or, since it often goes wrong, fails to do so.

Even if the mechanism did not quite fulfil its promise, the look of the Swissonic at least was a winner: akin to a 1970s bedside clock radio for the wrist (Derby, in fact, also made desk clocks) with all extraneous details kept to the bare minimum. Even the crown is situated at the rear of the case for added streamlining.

YEAR OF RELEASE 1974

MOVEMENT ESA 9176 electro-mechanical Dynotron

RELATIVE VALUE ★ ★ ★

NOTABLE FEATURE This watch has a rare jump minute mechanism

YEAR OF RELEASE 1972

MOVEMENT Valjoux 7734

RELATIVE VALUE ★ ★ ★

NOTABLE FEATURE Designed so the display is visible beyond the edge of a shirt cuff

46 DESOTOS 'CUFFBUSTER' CHRONOGRAPH

DESOTOS 'CUFFBUSTER' CHRONOGRAPH

If you are racing a car or an aircraft, when keeping time could be a potentially fatal distraction, you need a watch fit for purpose. To this end the Swiss brand Desotos came up with its 'Cuffbuster', with a Valjoux 7734 movement and a bridge signed by Sicura (the same was used by Breitling too) and, more importantly, a case with the pushers arranged across its top line, which actually makes their operation somewhat easier than the usual side-on arrangement.

More than that, however, Desotos gave its model a curious asymmetric arrangement that allowed the case to sit partly over the back of the hand, free from being covered by the wearer's sleeve and so enabling easier access. It is an elegant design solution – albeit one the wearer probably needs given the outsized dimensions of this watch – to a problem that is not an obvious one. Gianni Agnelli, the don of the Fiat carmaking business and generally regarded as one of the twentieth century's most stylish men, famously wore his watches fastened over his shirt cuffs for precisely this reason. This subsequently came to be seen as a style signature, though the Desotos would have solved his problem. In 'Cuffbuster' the idea certainly gave the watch a fantastic nickname too.

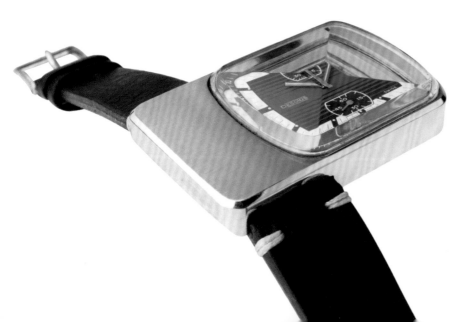

DUGENA CHRONOGRAPH

This chronograph was one of the results of a cooperative formed by the German section of the Swiss Union Horlogerie – led by Wilhelm Ulrich and Emile and Richard Rothman – and various watchmakers and watch retailers. The coop, established in 1917, was called the Deutsche Uhrmacher-Genossenschaft Alpina, more handily condensed into Dugena. The coop, the aim of which was to produce high-quality watches at a more accessible price, worked well, especially from the late 1940s, when, free from wartime production constraints and part of the new Federal Republic of Germany, its manufacture-to-shopfloor approach saw it become one of the most successful German watch brands of the next two decades.

Economic crises and various business decisions over the following years saw the brand's ownership shift – though it managed to weather the quartz crisis, being one of the few companies to quickly embrace rather than resist this new technology – and it is still operational, making mostly classic timepieces. Somewhat different from this model then, powered by the ubiquitous workhorse Valjoux 7734 movement, it is a distinctly 1970s period piece, with its oval shape, multiple colours on the dial and the unusual mounting of the pushers along the top line of the case.

YEAR OF RELEASE 1970s

MOVEMENT Valjoux 7734

RELATIVE VALUE ★ ★ ★

NOTABLE FEATURE Dugena was a manufacturer's cooperative focused on affordable watches

SPACE AGE DESIGN

If ever there was an object that had little logic in looking as if it could rocket through space, then an armchair might be it. Yet Steen Ostergaard's aptly named Meteor Lounge Chair, with its angular fibreglass shell and bold orange pads, was only one such piece of furniture produced during the 1960s. Nor was furniture alone in taking an out-of-this-world approach to the look of a product: cars, clothes, kitchen implements, graphics, toys and watches, not to mention art, television and film, all found inspiration in what seemed like the dominant theme of the zeitgeist: space travel.

It was, perhaps, a natural successor to the design of the early 1950s, inspired by the excitement surrounding the development of atomic energy. But ever since the Russians launched Sputnik, the first man-made satellite, in 1957, the so-called space race with the Americans – the race to achieve firsts in many aspects of space exploration, from the first spacewalk to the first landing on the moon – had shaped the visual culture. It was an expression of enthusiasm for visions of the future that promised such a clean, stark, ordered, prosperous and limitless contrast to the recent horrors of World War II, as well as societal upheavals such as the American civil rights movement and the Vietnam War.

Some objects experimented with the latest materials and with dramatic form to produce limited-edition products that suggested a new definition of luxury and sophistication. If you had the money, and lived in the West – or, more specifically the USA, where space age design was really championed – you aspired to live in a building akin to John Lautner's Elrod House in Coachella Valley, as featured in 1971's *Diamonds are Forever*, a movie itself featuring a plot about the development of a space-based laser weapon. You imagined sitting inside listening to David Bowie's 'Space Oddity' or watching your Keraclonic or Videosphere television – reminiscent of an astronaut's helmet.

You would wear a cut-out dress and helmet hat by André Courrèges – the first fashion designer to incorporate plastics and PVC into his collections – or something from Pierre

OPPOSITE The internal workings of Keracolor's B772 television may have been rather basic, but the complexity of its striking case – reminiscent, of course, of an astronaut's helmet – made it spectacularly expensive.

RIGHT Pierre Cardin's wool gabardine suit of 1966, with its cut-outs and felt helmet, nodded presciently towards the trouser suits worn by the flight attendants of Kubrick's *2001: A Space Odyssey*, two years later.

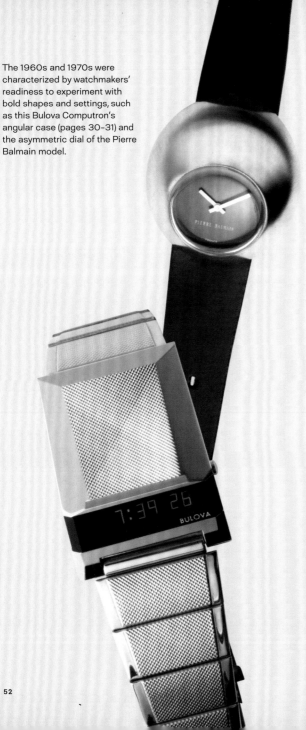

The 1960s and 1970s were characterized by watchmakers' readiness to experiment with bold shapes and settings, such as this Bulova Computron's angular case (pages 30–31) and the asymmetric dial of the Pierre Balmain model.

Cardin's Cosmocorps collection; or a minimalistic business suit by Hardy Amies, costume designer for Stanley Kubrick's *2001: A Space Odyssey* (1968). On your wrist would be one of André Le Marquand's Spaceman watches. You would perhaps wait excitedly for the likes of Ford's stunning Seattle-ite XXI concept car to become reality, but in the meantime certainly cherished what was parked outside: your Chrysler Turbine, the only car using jet technology as a means of propulsion that made it into (short-lived) production.

Yet space age design did not touch only the affluent. No matter how pedestrian the object, its design could provide a hint of tomorrow – for while the real space race was clunky and often deadly, its parallel in design took its cues more from science fiction, the space race as it might be (as it was somewhat optimistically estimated) just a few decades ahead. Even children were sold on this, thanks to the animated sitcom *The Jetsons*, about a future family living in a space colony, with their aerocar and robot maid.

This TV show was not the only thing to drive the escapist hype – certainly newspapers and magazines revelled in the possibilities seemingly afforded by this new age, not always with much truth behind the claims, creating a cosmic version of tulip mania. There was propagandistic value in this too: as the Cold War gained momentum, all this talk of space travel suggested – unlike the double-edged sword of the atomic age – a society fully in progressive mode, even if most of the space age buzz was ultimately about driving consumerism rather than real social change.

Indeed, as the space race fizzled out – the first moon landing, in 1969, seemed to announce some kind of winner, with the (to date) last moon landing happening only three years later – it came to be replaced with darker, more earthbound concerns, such as government corruption, energy crises, incoherent foreign policy and inflation. For its increasingly vocal naysayers, the space programme suddenly appeared to be a ludicrously expensive project.

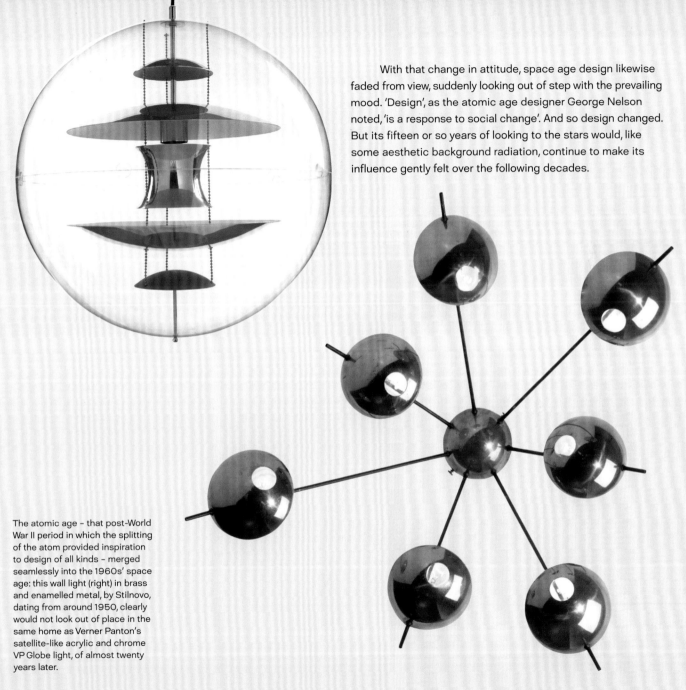

With that change in attitude, space age design likewise faded from view, suddenly looking out of step with the prevailing mood. 'Design', as the atomic age designer George Nelson noted, 'is a response to social change'. And so design changed. But its fifteen or so years of looking to the stars would, like some aesthetic background radiation, continue to make its influence gently felt over the following decades.

The atomic age – that post-World War II period in which the splitting of the atom provided inspiration to design of all kinds – merged seamlessly into the 1960s' space age: this wall light (right) in brass and enamelled metal, by Stilnovo, dating from around 1950, clearly would not look out of place in the same home as Verner Panton's satellite-like acrylic and chrome VP Globe light, of almost twenty years later.

DYNASTY MIDO SWISSONIC

Powering this stripped-back design is an ESA 9392 LED module, an in-house design of the ESA conglomerate, reportedly used in only two or three watch designs, among them those by the French brand Jaz. Outside this watch is all minimalistic – like many pieces of the 1970s, somewhat counter to the idea that the decade was aesthetically maximalist – with an undecorated case and just two side pushers, one to display the time, the other to show the date and seconds.

YEAR OF RELEASE 1975

MOVEMENT ESA 9392 light-emitting diode module

RELATIVE VALUE ★ ★ ★

NOTABLE FEATURE Minimalistic design in a maximalist era

ELGIN 'GOLF BALL'

With an over-case round and pitted like this, what else could this Elgin model be dubbed but the 'Golf Ball'? This was not a golf watch of the kind that some manufacturers have developed in more recent times, with a complication allowing a player to keep count of strokes and so forth, but it was certainly in high demand among golfers.

Indeed, American manufacturer Elgin was generally a popular brand at home thanks to the association of its Tank model with Elvis Presley (essentially the same 'Direct Reading' style watch in a different case) – he both wore the watch and gave it as a gift. Elgins were also known to have been worn by such historical figures as William 'Buffalo Bill' Cody, President Warren Harding and the baseball star Babe Ruth.

This watch's gold-plate pitted case perhaps afforded some protection from knocks while playing, although this was not its intention. The 'Golf Ball' came with a basic 17-jewel 717 calibre Elgin movement; the time was read from the two discs visible through the triangular window. This was perhaps meant as a visual nod to the shape of a driver golf club. Elgin promoted the fact that their watches were entirely American-made, though signed those watches with more advanced movements 'Lord Elgin'.

YEAR OF RELEASE 1957

MOVEMENT 717 calibre Elgin

RELATIVE VALUE ★ ★ ★

NOTABLE FEATURE The pitted dial cover mimics the surface of a golf ball

ENICAR 340
+ ENICAR SHERPA 320

If the watches of the early 1970s were enamoured of space age styling, actually looking as though they belonged some time in the future, these Enicar models keep the looks contemporary but play to the zeitgeist more, with that Saturn logo near the top of the dial. Indeed, it is the dials that really make these watches stand out, not just for their sunburst effect, but in the unusual colour choices of brushed golden brown and olive-gold (other versions were made with dials in equally distinctive shades such as indigo or ruby). Both featured the Enicar 1147B in-house automatic movement.

Enicar itself has a history dating to 1915, when Ariste Racine and his wife, Emma Blatt, launched Manufacture d'Horlogerie Ariste Racine in La Chaux-de-Fonds, Switzerland. Since Racine was part of a family with a watchmaking pedigree going back to the 1700s, the industry was littered with his namesakes, so the couple simply reversed the spelling of his surname to create Enicar.

What soon put the company on a firm financial footing was the creation of a model every bit as striking as the 340 or the Sherpa – a teardrop-shaped pocket watch with an aperture that could hold a compass or a photo.

But, in the postwar years, it was a reputation for precision and toughness – the company claimed that all of its movements were ultrasonically cleaned in the laboratory, leading to the creation of the Ultrasonic brand – that saw Enicar watches worn by adventurers of many kinds, notably a Swiss expedition to the Himalayan peaks Lhotse and Everest in 1956. It was this that, in the same year, inspired Enicar to name its explorer-type watches Sherpa, with more than a hundred variations of the Sherpa being produced over the following decades. There was a Sherpa Dive watch series too, its reliability demonstrated by fixing a Sherpa Ocean Pearl to the keel of the sailing ship *Mayflower II* on its 1957 Atlantic crossing.

YEAR OF RELEASE 1970s

MOVEMENT Enicar 1147B in-house automatic

RELATIVE VALUE ★ ★ ★

NOTABLE FEATURE Enicar was established by a husband and wife team in 1915

ETERNA DRIVER'S LED

Although the name Eterna first appeared on a watch in 1889, with the brand subsequently producing the first watch with an alarm in 1914, it is perhaps best known for its KonTiki model. That is named after the raft on which Thor Heyerdahl and five other Scandinavians in 1947 demonstrated that sea currents and Pacific winds could carry explorers from South America to Polynesia. All of the crew wore Eterna watches for this historic journey. The following year Eterna invented a movement with ball-bearing-assisted rotational movement, saving wear and tear on the parts and increasing the longevity of all watches it was fitted with: this advance is reflected in the choice of five indentations as Eterna's logo, seen here on this watch.

This, of course, is a long way from the kind of classic watch that Heyerdahl wore, and it is a reflection of the fact that, come the 1970s, Eterna was quick both to embrace electronic wristwatches with tuning fork resonators and to put quartz movements into analogue watches (while most companies associated quartz with digital display). This driver's model, with the display set at a side angle to the wrist, would likely not long survive the salt water Heyerdahl exposed his timepiece to, but it looked as though it could withstand maximum g-force on its way into space.

YEAR OF RELEASE 1970s

MOVEMENT Quartz digital

RELATIVE VALUE ★ ★ ★

NOTABLE FEATURE Eterna was one of the first brands to put quartz movements into analogue displays

FAVRE-LEUBA MOON RAIDER

As with so many watches of the 1970s, the name sought to capture the buzz around bold new ideas of tomorrow's world. But Favre-Leuba – creator of the first monopusher chronograph and the first mechanical aneroid barometer wristwatch – did not stop at the name, instead going out of its way to pile quirky design ideas into its automatic Moon Raider. There is an analogue display in a triangular dial, set in an asymmetric stainless-steel case with fibreglass sidings, with the date indicator set off-centre to the top of the dial, and with the crown set off-centre at the five o'clock mark.

The combined effect is to create a kind of arrowhead shape – an arrow pointing into the future perhaps. What would Abraham Favre have made of this? The watchmaker founded a family business that passed down through the line to the fourth generation, when it joined forces with Auguste Leuba. This was in 1815, making it one of Switzerland's oldest watchmakers.

YEAR OF RELEASE 1970s

MOVEMENT Automatic mechanical

RELATIVE VALUE ★ ★ ★

NOTABLE FEATURE Established in 1815, Favre is one of the oldest Swiss watchmakers

GIRARD-PERREGAUX CASQUETTE

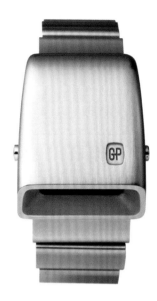

Typically considered a classic of 1970s design – not merely watch design – the Casquette is replete with innovation. Small wonder, when Girard-Perregaux, certainly an esteemed maker, created its own electronics department to develop this watch in 1974. It took two years of work, with Girard-Perregaux building up to the watch in stages, first creating an electronic clock that could control subordinate clock units, then a desktop clock in a housing made of Macrolon. This was a lightweight glass-fibre composite polymer that, in a watchmaking first, would eventually be used in one version of the Casquette, although it proved less efficient when shielding the components inside from the negative effects of ambient electromagnetism. Other versions of the watch came in 18k rolled gold or – the most popular version, of which some 4,000 were made – in stainless steel.

Atypically, Girard-Perregaux's decision to develop its own LED module, rather than buy one in from the popular makers such as ESA, National Semiconductor or Hughes Aircraft, made it an expensive project and consequently an expensive watch. In fact, for one calibre – there were three in all, each advancing on the previous design – the semiconductors were mounted on flexible boards, likely a world first. Power consumption was high too, which is why consumers had to pick what they wanted. One option was to have a twenty-four-hour display, with the date but not the month; another was to have time and date showing brightly, but seconds more dimly. In the end only 8,000 Casquettes were made in all, on the market for just a two-year period.

Yet the result is, for many collectors, the most pure of forms among all driver's LED display watches of the 1970s, or indeed since: from the elegant tapered form of the solid case to the stylish branding and the chunky, tactile links of the bracelet (a leather strap was also available). Press the button on the left to set time and date, the one on the right to display the time, another push for the date, another for running seconds. It looked great sleeping too.

YEAR OF RELEASE 1976

MOVEMENT Girard-Perregaux calibre 395 quartz

RELATIVE VALUE ★ ★ ★

NOTABLE FEATURE One Casquette model was the first to use a case of glass-fibre polymer

GUBELIN
G-QUARTZ CUFF

If the 1970s were liberating years for men's style, as a touch of feminine flamboyance became more widely accepted (at least in metropolitan circles), then why not dispense with traditional leather straps or clunky bracelets and embrace the cuff instead? Gubelin, a Swiss international jewelry retailer established as a watchmaker's shop in Lucerne in 1854, came up with its G-Quartz design – or rather, an 'exquisite design coupled with the most advanced precision', as a print advertisement of the time had it. Here was a rather simple round dial mounted in a square case (housing an ETA 9181 movement) and then – most distinctively – set into a stainless-steel open bangle. This could be gently bent to fit any wrist, assuming it was modernist enough in spirit to take this unusual design.

YEAR OF RELEASE 1973

MOVEMENT Quartz ETA 9181

RELATIVE VALUE ★ ★ ★

NOTABLE FEATURE The bangle could be bent to better fit most wrists

HAMILTON DATELINE
TM-5903

YEAR OF RELEASE 1969

MOVEMENT Buren-made micro-rotor automatic

RELATIVE VALUE ★ ★ ★

NOTABLE FEATURE Hamilton's first model made outside of the USA

Although an American company, founded in Pennsylvania in 1892 and associated with a long history of American watchmaking, in 1969 Hamilton would move much of its production to Switzerland, in large part through the purchase of Buren, doubling capacity in the process (or at least until 1972, when capacity shrank and the Buren-Hamilton factory was liquidated). This Dateline model was one of the first to be made abroad, complete with Buren's Intramatic micro-motor automatic movement. This integrated the rotor at the same level as the rest of the movement and so allowed the case to be unusually thin, an idea Hamilton marketed under the Thin-o-Matic name (TM stands for this).

Hamilton had made models under the Dateline name since 1954, when, for a decade or so, a date complication was a relatively rare idea. The A-677 Dateline model, which ceased production in 1969, was considerably more classical than the model pictured, which marked a sea-change in styling: the date window is given greater prominence, the dial is grey and the bold cream horizontal indices are clearly more decorative than functional. The 1971 version was also available on a woven metal bracelet.

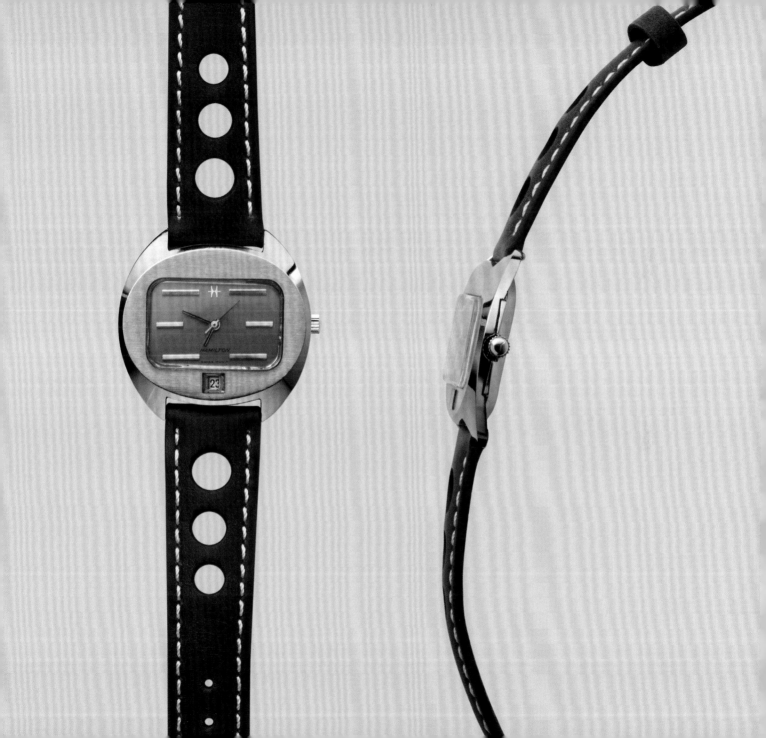

As it says on the dial, this Hamilton model is electric – and small wonder that the company was keen to drive this point home. In 1957, at the Savoy Plaza Hotel in New York, Hamilton had unveiled the first mass-production 'moving coil' electric movement, known as the 500. It was a technology the company's lead engineers Arthur Fillinger and Fred Koehler had been working on for a decade, coming up with an oscillator system using electrical current that could work on the wrist, and a platinum and cobalt battery small enough to provide that current. Hamilton had to approach more than forty battery makers before one, the National Carbon Company, agreed to work with it on the project.

The result was, Hamilton claimed, 'the first basic improvement in 477 years of watchmaking' (referring to Peter Henlein's probable invention of the watch in the fifteenth century). The 500 went into several Hamilton watches, including the Victor. The electric pieces were not without their running troubles,

HAMILTON ELECTRIC VICTOR

nor their commercial ones, not least jewellers' reluctance to sell them given Hamilton's recommendation that all service work be handled by its factory – a mistake that rival Bulova would not make when it introduced the Accutron three years later. But then the advent of quartz-regulated movements in 1969 would soon put paid to electric watches altogether anyway...

Still, this is not the only story behind the Victor. The dial's gold-on-black sunburst decoration, round indices and nods to Art Deco, and the asymmetric squared case with the crown on one corner, brought together the ideas of American industrial designer Richard Arbib (1917–1995). He is better known as an automotive designer for General Motors, American Motors Corporation and Packard, including the latter's Pan American model, and was one of the designers to popularize the idea of putting fins on cars. His Astra-Gnome concept, a 1956 vision of what the car might look like in 2000, included a Hamilton 'celestial time-zone clock permitting actual flight-type navigation', as a no doubt perplexed *Popular Science* cited the designer as saying.

Indeed, Arbib would design more than cars, including cameras for Argus, boats for Century, jewelry for Swank and watches for Benrus as well as Hamilton. For the latter he designed not just the Victor but also the Everest and Pacer models, and the triangular-cased, finlike Ventura – another watch made famous by association with Elvis Presley and, latterly, the *Men in Black* movie series. Hamilton gave Arbib an entirely open brief – and he ran with it.

YEAR OF RELEASE 1957

MOVEMENT Moving coil electric

RELATIVE VALUE ★ ★ ★

NOTABLE FEATURE One of the first Hamilton designs by Richard Arbib

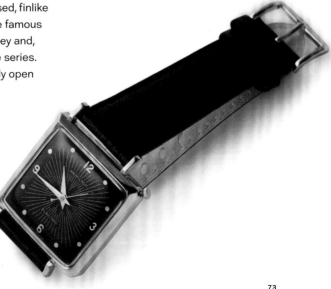

HAMILTON ELECTRONIC

Before quartz really changed everything in the watch industry, making many technologies obsolete, including, almost, traditional mechanical movements, there was still room for something like the Dynotron ESA 9158 (or Hamilton calibre 702) in this electronic watch. The movement is something of a hybrid, a mechanical balance and mainspring, with lever adjustment, which instead interacts with a transistor oscillator. But this Hamilton Electronic wins as much on style, with its angular case giving the appearance of solidity, and the uncluttered dial a hint of the future.

YEAR OF RELEASE 1970

MOVEMENT Hamilton calibre 702

RELATIVE VALUE ★ ★ ★

NOTABLE FEATURE The movement was a rare mechanical/transistor hybrid

HAMILTON FONTAINEBLEAU B
+ HAMILTON FONTAINEBLEAU CHRONO-MATIC

The Fontainebleau B's predecessor, the Fontainebleau, was so named after the Palace of Fontainebleau, near Paris, where an oval courtyard inspired designer Ulrich Nydegger to come up with a patented case shape for the watch in 1968. It is unclear whether a much more commonplace square courtyard likewise inspired the outsized, day/date B model (with Swiss self-winding 64A automatic movement).

It was, however, promoted as being waterproof, making it one of the first square watches to be so – those right angles presenting a technical challenge to achieving watertightness – alongside the more famous Heuer Monaco, which was launched around the same time; indeed, Heuer built the case for Hamilton as part of Project 99, of which more below. Like the Hamilton Dateline, the Fontainebleau benefits from those unusual horizontal indices and makes an eye-catching graphic device of stacking the day/date indicators rather than placing them, as is much more common, side by side.

The third and final of Hamilton's Fontainebleau series was the Chrono-Matic version, which came with the Hamilton calibre 11, developed as a joint venture called Project 99 with Breitling and Heuer, ostensibly because sales of automatic watches were cutting deep into the latter's own chronograph sales. The result was arguably the first ever automatic chronograph movement – and certainly Hamilton trumpeted the fact in its ads, speaking of its Chrono-Matic as 'an outstanding technical achievement due to co-operation between three famous Swiss watchmakers'.

It was, at least, the first ever chronograph module with automatic winding. Zenith has long argued that it created the first fully integrated chronograph automatic-winding movement, driving the point home by naming it the El Primero – yet Seiko may have beaten even them to this in Japan by a few months.

The Fontainebleau Chrono-Matic returned the series to Nydegger's 'squared cambered' case design

YEAR OF RELEASE 1969

MOVEMENT Swiss self-winding 64A automatic

RELATIVE VALUE ★ ★ ★

NOTABLE FEATURE The Fontainebleau was inspired by the shape of a palace courtyard

(although Hamilton would also produce non-Fontainebleau Chrono-Matics in more traditional round cases). Perhaps appreciating that the case shape was statement enough, Hamilton kept the dial details to the minimum, using only three sober colours – black, grey and white – with the sub-dials coloured panda-style, black on white. When the watch was first unveiled it was shown with more distinctive paddle hands, but these seem to have been replaced in favour of more commonplace sword-style hands in the production version.

A tell-tale sign of the movement, which would be found in watches from co-developers Heuer and Breitling too, is that the crown is situated on the left side of the watch, with the pushers, meanwhile, on the right. It was figured that with an automatic movement, winding or time-setting would only be very occasionally required. After all, their wearers were probably too busy to do so. 'Men of action do not wear watches, they wear chronomatics!' as a 1970s Heuer ad put it.

YEAR OF RELEASE 1969

MOVEMENT Hamilton Calibre 11

RELATIVE VALUE ★ ★ ★

NOTABLE FEATURE Designer Ulrich Nydegger's 'squared cambered' case shape

HAMILTON ODYSSEE 2001

It is a shame that the watch Hamilton was commissioned in 1966 to devise for Stanley Kubrick's *2001: A Space Odyssey* (1968) never went into production. With its multi-dialled front, outsized crown and integrated plastic-rubber bracelet, it would have been a welcome addition to the pantheon of intriguing watches designed throughout the 1960s and 1970s, had its complexity not prevented it from getting beyond the prototype stage and actually to market, as Hamilton had planned.

But it did, at least, inspire Hamilton to produce its commemorative Odyssee 2001 watches instead – the spelling changed to the French for copyright reasons. Neither looked anything like the watch proposed for the movie. But both came with Hamilton's 694A automatic movement, in a round case, with integrated crown and dial set slightly off-centre – sloping towards the wearer for easier reading – with a mesh bracelet and suitably space age detailing. The rarer and more striking of the two versions came with bold triangular hands and indented indices. Kubrick might have proved the ideal customer for these: he always wore two watches, one on each wrist.

In 2006 Hamilton did, in fact, finally release a close working version of the *Space Odyssey* prop, with the main dial powered by a mechanical ETA movement and each of three sub-dials driven by a quartz movement. It was, of course, limited to a run of 2,001 pieces.

YEAR OF RELEASE 1968

MOVEMENT Hamilton 694A automatic

RELATIVE VALUE ★ ★ ★

NOTABLE FEATURE Inspired by Hamilton's prop design for *2001: A Space Odyssey*.

HAMILTON ODYSSEE 2001

HAMILTON QED LED

'QED' stands for 'quartz electronic digital', and this model was one of a number based on technology developed by Pulsar but branded by Hamilton. The tech afforded the watch a complex (for the time) fifteen-function split-seconds stopwatch, but just as impressive is the one-piece integrated case and bracelet.

YEAR OF RELEASE 1974

MOVEMENT Hamilton QED solid state quartz electronic

RELATIVE VALUE ★ ★ ★

NOTABLE FEATURE QED stands for 'quartz electronic digital'

HAMILTON SELF-WINDING

One of a number of watches Hamilton produced branded 'self-winding', with a mechanical automatic movement, this watch is unusual for its red second/minute markers and matching red second hand.

YEAR OF RELEASE 1970

MOVEMENT Hamilton self-winding mechanical automatic

RELATIVE VALUE ★ ★ ★

NOTABLE FEATURE Red second/minute markers and matching second hand

HAMILTON THIN-O-MATIC T-403 'SHARK FIN'

This is another watch designed for Hamilton by the industrial designer Richard Arbib, the distinctive pale wedge on the two-tone dial and the asymmetric lug casings playing to his love of fins, a design detail he more typically applied to cars. This came with a calibre 663 automatic movement with micro-motor, making it one of the few non-electric watches Arbib created for the American manufacturer. Of the some seventy Thin-o-Matic watches made by Hamilton, this is among the most elegant.

YEAR OF RELEASE 1960

MOVEMENT Hamilton calibre 663 automatic with micro-rotor

RELATIVE VALUE ★ ★ ★

NOTABLE FEATURE One of the few non-electric watches by Richard Arbib

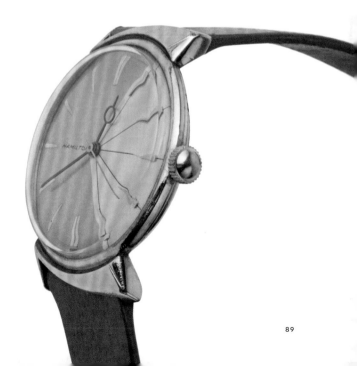

90 HANDCUFF BRACELET WATCH

HANDCUFF BRACELET WATCH

The origins of this watch are unknown, but it points to the possibilities in form afforded by the new accessibility of moulded thermoplastics during the mid-1960s. Their affordability (and toughness) also introduced the idea of fun to watch design – since one was no longer buying a watch for life but building a watch wardrobe – hence the novelty of this piece, with its ratchet-to-lock bracelet design. Plastics ushered in an era of experimentation in shape, flexibility, colour, lighter weights and varying degrees of opacity to watch cases and straps.

YEAR OF RELEASE 1960s

MOVEMENT Swiss manual winding

RELATIVE VALUE ★ ★ ★

NOTABLE FEATURE Plastics allowed for lighter, cheaper, more expressive designs

HEUER FORD RS SPLIT LAP UNIT 77 LCD CHRONOGRAPH

If ever a watch were to appear on the wrist of a character in some dystopian sci-fi film and look the part, it would likely be this one. The Heuer Ford RS Split Lap Unit 77 was, atypically for the time, designed not by its manufacturer (Heuer, which later amalgamated to become the better-known Tag Heuer) but by the engineers at Ford Rallye Sport (RS). It was then manufactured by Heuer, based around existing chronographs, notably the Kentucky and the Senator. The RS Split Lap, of which around 3,000 were made, is striking for its two-display arrangement, similar to Heuer's Chronosplit model, and for the angular, ergonomic fit on the wrist, which – like most driver's watches – sets the time display towards the wearer's viewpoint when hands are on the wheel.

The most futuristic aspect of the design, however, is the two round elements at the lower end of the watch. These are actually the watch's battery apertures. Showing the insides on the outside, as it were, might have been inspired by Richard Rogers's Pompidou Centre in Paris, which happened to open the same year this watch was launched; in practical terms, it made changing the batteries that much easier than most other watches. Curiously, the Heuer name does not appear anywhere on the watch: Ford's name appears in the technologically advanced upper liquid-crystal display window and on the stainless-steel bracelet's clasp. Ford advertised the watch alongside its rally car of the time, the Escort RS2000, which the creation of the timepiece was to promote. The watch has not dated as badly as the car.

YEAR OF RELEASE 1977

MOVEMENT Quartz calibre 103

RELATIVE VALUE ★ ★ ★

NOTABLE FEATURE Designed by Heuer but unsigned

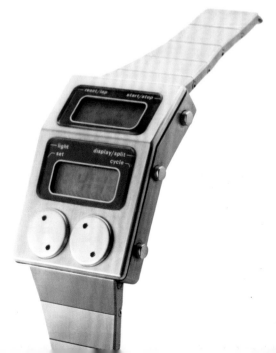

HUDSON DIRECTIME + ITRACO DIRECTIME

Mechanical digital display watches, sometimes called digit wheel or jump hour watches, were one way in which typically smaller, more nimble watch brands of the 1970s were able to respond to the onslaught of quartz movements out of Asia. What appeared to be a category killer – why have a mechanical watch, fragile and in need of winding and servicing, when you could have a much more accurate, battery-powered watch using the technology of the future? – proved able to be resisted, at least for a while, by aping the display.

That is what Hudson did with this Directime manual-winding model. This offered the jump hour style, with the new hour jumping into view as the sixtieth minute passed, though other watches of the period saw the hour indicator slowly turn with the minutes, often leaving it showing awkwardly somewhere between the hours. The Directime's case gave a suitably timely nod to the space race theme of much design of the era.

The Hudson Watch Company was Swiss but essentially created for sales to the US market. It was registered by Antoine Castelberg in La Chaux-de-Fonds and New York in 1884. Hudson often produced its models under other brand names, including Gisa, Carlton, Adelphi, Globe and Itraco – like the one pictured, with its integrated case/bracelet – among the many brands of the time whose history is much less clear today.

YEAR OF RELEASE 1971

MOVEMENT Swiss manual winding

RELATIVE VALUE ★ ★ ★

NOTABLE FEATURE Jump hour display

YEAR OF RELEASE 1970s

MOVEMENT Swiss mechanical winding

RELATIVE VALUE ★ ★ ★

NOTABLE FEATURE Itaco was a spin-off brand created by the Hudson Watch Company

THE RETRO AESTHETIC

The past is a foreign country, as L. P. Hartley wrote: they do things differently there. They certainly designed them differently too. Indeed, if design is more typically associated with the idea of creating something new – offering a new solution to a pressing problem, or devising an original, maybe even shocking aesthetic – the twenty-first century has also seen the discipline keen to look back in time for inspiration.

From cars to fashion, furniture to interior design, graphics to watches, lazily (as some might say) reissuing a design from the archives, or taking a modern product but making it look, at least superficially, as though it belongs to a past era, has become a recognized design style. It is typically dubbed – not always in a positive way – 'retro' or 'heritage'.

In parallel to this has come an ever-increasing appreciation for the actual products of the past: if once the old was shunned – suggestive as age is to some of inefficiency and outmodedness – 'vintage' has become an adjective denoting the cool and collectible, with all the added status inherent in a product that is both rare and the real deal, rather than an imitation. And this even given the hassles of using a product that is outmoded too, be that a vinyl record, 'old school' games console or, yes, a mechanical watch.

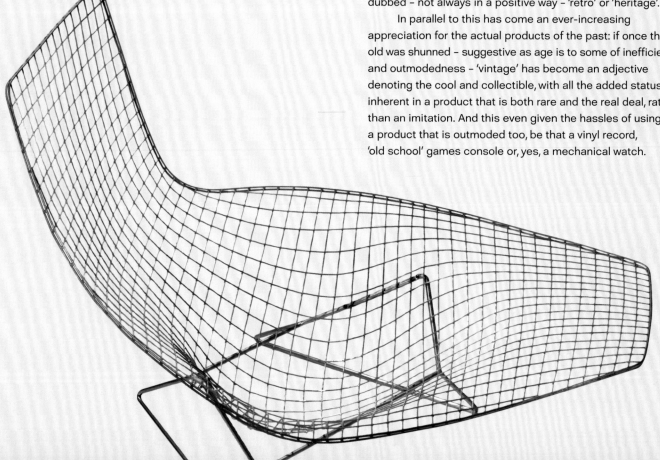

Retro is not a new idea, of course. In architecture, neo-classicism, with its visual references to ancient Greece and Rome, was a popular style of the 1790s. The clothing fashions of the 1960s more than nodded to the 1920s. And so on. But retro does appear to have a particularly post-millennial appeal. In part, it might be argued, the aesthetic of the past offers a sense of familiarity, especially to those who never lived through the times to which the aesthetic refers; for those who have, the design might merely look dated or regurgitated. Certainly retro often ends up offering a kind of hyper-real, sometimes cartoonish version of the past, one fuelled by a nostalgia more imagined than real.

Yet, all the same, this familiarity is reassuring, all the more so in troubled times, or when consumerism offers competing choices that might feel both overwhelming and immediate. Retro is a braking mechanism. We live, in Douglas Coupland's phrase, in an accelerated culture. The Internet age means that everything happens everywhere almost instantaneously, with everyone everywhere knowing about it. The new, or what passes for it, does not have the opportunity to take root or explain itself before it has been trumped by something newer vying for our attention. As it is happening, history always looks messy and confusing, but the space to look back gives clarity again. Retro is the opportunity to regain perspective and make life tangible, as well as to step back from the global homogeneity that the corporate world has championed. Whereas we once revolted by generating the innovative, challenging or even confrontational – the stuff of the future – now we revolt to the past.

In that, we find an authenticity so often lacking in the new. If the current can feel fresh, it can also feel disposable and without foundation, which is why brands that have an impressive history are always keen to underscore the fact. The products of the past – both genuine and ersatz ones – gain credibility in seemingly having weathered the passage of time. They have a look that make them close cousins to that other tribute to history in design: that of being a

THE RETRO AESTHETIC 99

'classic'. If a classic is often considered so because in some way it has managed to transcend the era in which it was created – it is literally timeless – a retro object is held in high esteem by more obviously originating in a particular era and embodying the mood of those times too.

That works especially well if, largely thanks to the way it has been mythologized and kept alive by the media, and through film especially, that era has come to be regarded as particularly cool. And, while every past decade seems to have its hardcore fan base, few decades are quite as cool as the 1950s, 1960s and 1970s, with their blend of James Dean, Carnaby Street, Woodstock and Studio 54.

It has been argued that this cool is, in part, determined by those whose age make them part of the dominant creative and consumer groups, who tend to fetishize the era of their childhood and its preceding decades: so, the theory goes, this perceived cool shifts up through the decades as younger people get older and find themselves with both influence and disposable income. Retro has a slowly moving focus.

But it is also tempting to argue – perhaps from the bias that comes with one's own age – that the period of the postwar consumer boom actually was particularly cool. Not only was there a proliferation of products – with the watch industry, for example, then sustaining multiple small companies – but there came too the realization that attractive design in itself was a selling point for mass-market or commodity items; that a watch by one of these small companies could find a market simply by being really good-looking or progressive, qualities sought decades later.

Therein lies a certain irony, of course. The products and their styles now appreciated as retro typically aimed at being anything but nostalgic when they were first used: rather they aimed at being cutting-edge, even futuristic. Perhaps this is really why so many of the watches in this book still feel so modern, even while belonging to the past. They are retro, but they also feel utterly relevant.

PAGE 98 Painter and sculptor Harry Bertoia is just as well known for the design of his classic Bertoia Diamond and Side chairs (1952) for Knoll, with upholstery laid over latticed steel – this being an early prototype.

PAGE 99 TOP Ball clocks – like this 4755 model from the Howard Miller Clock Company, launched 1948 – nodded to the postwar era's fascination with the new world seemingly offered by leaps in chemistry and physics; today they chime with a more pop sensibility.

PAGE 99 BOTTOM Introduced in 1962, the Braun Station T1000 CD radio (CD referring to the German diplomatic corps, not compact disc) was another classic design from Dieter Rams, whose functional aesthetic still informs much design today.

LEFT Finnish designer Eero Saarinen's Tulip chair for Knoll, dating to 1956, helped give industrial materials such as fibreglass and aluminium a new luxury appeal.

The advent of 'total' lifestyle consumerism in the 1950s and 1960s began to see an integration of design aesthetics across various products. Here, Fritz Hansen's Egg leather and plastic shell chair of 1958, Serge Mouille's spidery light fixture of 1953 and Lip's much later Mach 2000 watch of 1975 (pages 124–27) share a taste for the monochrome and minimalistic that resurfaced in the 1980s and 1990s.

102 JAEGER-LECOULTRE 'DISCO VOLANTE'

JAEGER-LECOULTRE 'DISCO VOLANTE'

YEAR OF RELEASE 1970

MOVEMENT Ref. 562 42 Swiss calibre K883 JLC automatic

RELATIVE VALUE ★ ★ ★

NOTABLE FEATURE The 'disco volante' case shape takes its name from 'flying saucer' in Italian

This automatic steel model from Jaeger-LeCoultre is exemplary of the unconventional 'disco volante', or 'flying saucer', case shape popular through the 1950s and into the early 1970s. It was characterized primarily by the atypically wide – and sometimes asymmetric – bezel, one that, viewed from above, often entirely hid sight of both the crown and the lugs, creating a modernist streamlined look in the process.

Initially nicknamed by Italian collectors, the popularity of disco volante models naturally echoed the fascination with alien visitors that peaked over the same period. However, disco volante models actually pre-date the buzz that popular culture brought to that subject, some models dating to the 1940s. At its height many high-end brands, including Omega, Audemars Piguet, Rolex and Patek Philippe, also produced disco volante models, albeit with vastly different ideas as to what width of bezel was enough to warrant the name.

JAEGER-LECOULTURE MASTER-QUARTZ MODELS

YEAR OF RELEASE 1972

MOVEMENT Jaeger-LeCoultre quartz calibre 352

RELATIVE VALUE ★ ★ ★

NOTABLE FEATURE The movement was developed as a side project to the Beta 21

If bolder design tends to be associated with smaller brands, with the globally known stalwarts of the Swiss industry historically leaning towards more classical design, that has not always been the case – at least when it comes to those subtle details that can make a watch a hit or a miss. Both of these Jaeger-LeCoultre Master-Quartz models, for example, exhibit a design trait more typical of watches of the 1960s and 1970s: a playfulness with the design of the minute indices, here with one model using a pattern of blocks of graduated size and shade, the other an unusual pendant formation of fine lines.

Many upscale watches of the era used the Beta 21, a quartz movement developed collaboratively in the mid-1960s by Swiss movement makers, spearheaded by Girard-Perregaux, in a bid to compete with the influx of quartz movements from Japan. The result, however, was on the clunky, chunky side, so Girard-Perregaux ran a side project to develop its own quartz calibre, the 352.

That is what powers these Master-Quartz models. With additional funding from Jaeger-LeCoultre, the hybrid-like movement was developed by Girard-Perregaux's Georges Vuffray and went into production in 1971, forming one of Jaeger-LeCoultre's earliest forays into quartz movements. The movement comprises the mechanical element, stepping motor, electronic module and, of course, the chip with quartz crystal, which was supplied by Motorola.

Here was a model that bridged two worlds, the practicality of electronic technology combined with the quality standards of haute horlogerie. This, of course, was not a combination with much of a future: quartz movements were soon mass-produced and ubiquitous, with an air of disposability about them, so that putting them in a housing of excellent quality began to look nonsensical. Such a housing required a movement of comparable craft; the question was whether anyone still wanted such an object.

JOVIAL VISION 2000

'You may not be able to go to the moon yet but you can already wear this Vision 2000 watch of outer space concept,' ran the print advertisement for this watch. It pictured some kind of alien planet surface (a long way from the images of the real lunar surface provided by the two or three moon landings that had occurred by this time), together with the watch itself standing there silently, an obvious nod to the monolith of Kubrick's hugely influential *2001: A Space Odyssey* (1968). Indeed, there is something to be said for the idea that the designs of such 'space man' watches owed a debt more to the big-screen science fiction of the era than to the actual space race.

The aesthetic of this watch, however, almost justifies such pomposity. This sideview design, with Swiss manual-winding mechanism, certainly had an extraterrestrial quality to it, with the space helmet case of stainless steel distorting anything reflected in it. As another ad had it, here was 'the watch that tells the time with tomorrow's styling'. It would certainly have struck Fernand Droz as out of this world – he was the founder of Jovial, in La Chaux-de-Fonds in 1929, four years before H. G. Wells published his speculative science fiction novel *The Shape of Things to Come*.

YEAR OF RELEASE 1970

MOVEMENT Swiss manual winding

RELATIVE VALUE ★ ★ ★

NOTABLE FEATURE This was the watch 'that tells the time with tomorrow's styling'

JULES JÜRGENSEN WEDGE

Is there something Danish about this watch's design, with its warm colours – the bronzed case – and simple, functional shape, the wedge of the case angling the watch gently towards the wearer's gaze? That may be reading too much into it but, all the same, it was a Dane, Jules Jürgensen – son of pioneering Danish watchmaker Urban Jürgensen and grandson of one of Denmark's oldest watchmakers – who trained as a watchmaker in Geneva and, in 1835, established a watchmaking business there with his own sons.

The business patented a hand-setting mechanism and won multiple awards, and its pocket watches and marine chronometers enjoyed

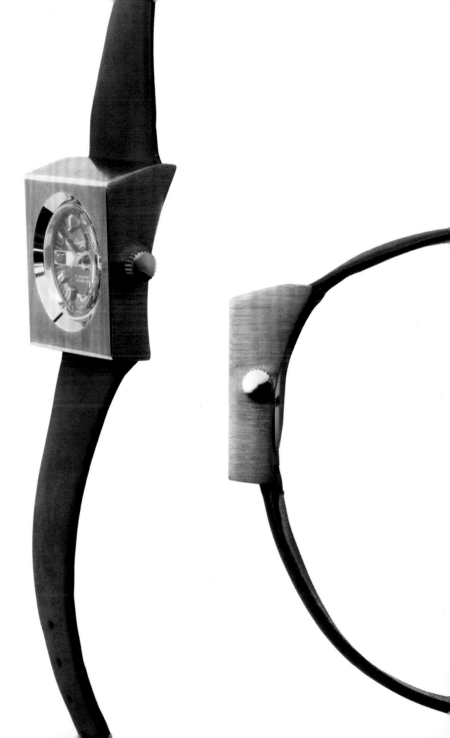

such a reputation back in Denmark that fellow Dane Hans Christian Andersen once paid it a visit when passing through Switzerland. Andy Warhol owned a Jules Jürgensen watch.

It remained a family firm until 1916, when it was bought, and in 1919 began to make wristwatches. In 1936 it was acquired by a New York company by the name of Aisenstein-Woronock. Soon after this mechanical manual-wind watch was made, the company was acquired again, in 1974, when the brand began to use quartz movements to reposition Jules Jürgensen as more a producer of value watches, leaving the Urban Jürgensen name to focus on haute horlogerie pieces.

YEAR OF RELEASE 1970

MOVEMENT Swiss manual-winding mechanical

RELATIVE VALUE ★ ★ ★

NOTABLE FEATURE The Jürgensen name is historic in Danish watchmaking

LECOULTRE HPG MEMOVOX

YEAR OF RELEASE 1970

MOVEMENT Automatic calibre 916. ref. E3072

RELATIVE VALUE ★ ★ ★

NOTABLE FEATURE HPG stood for 'High Precision Guarantee'

The Memovox – 'the voice of memory' – was an apt name for this LeCoultre watch. It had an automatic calibre 916 movement, designed in 1969 and known as the 'Speedbeat', of such accuracy that it was in service in various models for the next twenty years; sales in America, at least, saw the likes of this watch sub-branded HPG, for 'High Precision Guarantee'. But the watch also had a manual-winding mechanical alarm, chiming to remind the wearer of some or other important appointment. Hence the two crowns.

Watchmaker Jaeger-LeCoultre (which, until 1980, sold most of its watches under the shortened LeCoultre name) had been producing alarm watches since 1950, the intention being to compete with Vulcain's pioneering Cricket alarm watch, launched the previous year. But Jaeger-LeCoultre took a different path from Vulcain, separating the power source for timekeeping and alarm functions. In 1956 it created the world's first automatic alarm watch. Indeed, the company somewhat ran with the idea, towards the end of the decade producing a Deep Sea Alarm Automatic – a diving watch with an alarm – and the rather prosaically named Memovox Parking, a watch to remind its wearer to feed the parking meter.

YEAR OF RELEASE 1969

MOVEMENT Automatic calibre AS 1902

RELATIVE VALUE ★ ★ ★

NOTABLE FEATURE Designed by illustrator Prince François de Baschmakoff

LIP BASCHMAKOFF JUMP HOUR

Not many watches can claim to have been designed by even minor royalty. But the Baschmakoff Jump Hour was just one of a number of watches designed for the innovative French watch company Lip over the mid-1970s by one Prince François de Baschmakoff.

He was Lip's first freelance designer and arguably brought a bold new direction to the brand's aesthetic as a consequence of not being a specialist in watch design. Rather, de Baschmakoff was a Paris Ecole des Beaux-Arts-trained illustrator and colourist, working for fashion magazines and designing packaging for the likes of French *grand magasin* Printemps; indeed, one award-winning packaging design of his was for a globe-shaped plastic box he created for Lip. His breakthrough with that company, however, came in 1968, when he filed an application for a French patent for a mechanical watch with a digital time indication and showed the idea to Fred Lip, Lip's founder.

This particular automatic mechanical model required the development of a special hairspring to gather the necessary power to make the jump hour jump: the watch comes with a spring-lever for stopping the hour disc in the right horizontal position when the ratcheted edge of the disc reaches the lever. In order to recoup the cost of development and manufacture it was likely that this design appeared under a number of guises and under different brands, among them Vulcain, Hoga and Damas. But such was the radicalism of the design that it opened the door for further pioneering independent watch designers such as Isabelle Hebey and Roger Tallon to design for Lip too.

LIP BY ISABELLE HEBEY
+ MISCELLANEOUS MECHANICAL LIP MODELS

The indices may have subtle variations, but it is the asymmetric dials, simple case shapes in anodized or chromium-plated alloy, articulated lugs and gradually narrowing straps that make these T-13 calibre manual-wind watches so distinctive. All three – from a series of eight – are the work of Isabelle Hebey, one of a group of seven industrial designers hired by Lip through the early 1970s to lend their eye to watch design.

That was the visionary, if perhaps too late, decision of Claude Neuschwander, successor in 1971 to Lip's founder Fred Lip, and the man who appointed illustrator and packaging designer François de Baschmakoff to overhaul the Lip brand. He did so in part by bringing in another six designers with no experience in watch design, specifically Roger Tallon (who would, arguably, have the greatest impact on the brand), Michel Boyer, Michel Kinn, Rudi Mayer, Marc Held and Hebey.

As strong as it was, the style these designers brought in was more a continuation of that which Fred Lip had been gradually introducing over the 1960s. This is perhaps exemplified by the gear change evident between the super-minimalistic yet nonetheless classical R 23 mechanical Lip watch pictured on pages 122–23 – released in 1960 – and the asymmetric model with mechanical Peseux 320 manual-wind movement, released just a year later (pages 120–21). Yet certainly Hebey's watch designs took this line and ran with it.

Hebey began working in Paris during the 1950s, designing everything from public housing to pens, cars – such as the Honda Accord, as well as Honda's logo – and car interiors, and homes for celebrities including Sophia Loren and Yves Saint Laurent, pioneering along the way an aesthetic of mixing the antique with the cutting-edge. She had a thing for lining walls with steel sheeting, something perhaps reflected in the matte cases she designed for these watches.

In 1965 Hebey said that she understood her clients' fear of de-cluttering because she too shared 'the passion for useless objects'. Did she presciently count mechanical watches among those? After all, amid competing technologies and employment strife, including an eight-month-long worker-led occupation of the factory, Lip would be liquidated in 1977, just three years after the release of Hebey's watches shown here.

YEAR OF RELEASE 1973

MOVEMENT Mechanical-winding LIP T-13 calibre

RELATIVE VALUE ★ ★ ★

NOTABLE FEATURE Isabelle Hebey also designed the interior of Concorde

YEAR OF RELEASE 1970

MOVEMENT Mechanical Duromat 451 R movement by Durowe

RELATIVE VALUE ★ ★ ★

NOTABLE FEATURE Clean minimal dial and modern sporty design

YEAR OF RELEASE 1961

MOVEMENT Mechanical Peseux 320 manual winding

RELATIVE VALUE ★ ★ ★

NOTABLE FEATURE Daring asymmetrical case design and very modern dial layout

YEAR OF RELEASE 1960

MOVEMENT Mechanical LIP R 23

RELATIVE VALUE ★ ★ ★

NOTABLE FEATURE Extreme minimalism contrasts with later designs as introduced by Hebey

LIP MACH 2000 CHRONOGRAPH + LIP MACH 2000

Arguably the most famous of all creations from the French watchmaker Lip, established in 1867, the Lip Mach 2000 was as forward-looking as its supersonic name implied. The watch was created by Frenchman Roger Tallon, a designer who turned his hand to designing the revolutionary Téléavia P111 portable television, the modular, space age Model M400 Helicoid Spiral Staircase, the Derny Taon 125 motorcycle, the 8 mm Duplex camera, the Wimpy chair and the TGV Atlantique bullet train (just one of the many trains he designed), among much else in his varied portfolio. Each design speaks to the fact that Tallon trained as an engineer.

As with much of Tallon's design work, the Lip 2000 embodied a certain deliberate playfulness absent in most watches of the time, not just in the offsetting of the dial relative to the strap, but in the blunt-ended hands, and most of all in those ball-shaped, primary-coloured push-and-pull buttons, reminiscent of the Eames's 1950s coat hooks. Each function is colour-coded. Their joyfulness is somewhat at odds with one Lip 2000 model name: Dark Master. The original versions of these designs, as pictured, came with a manual-wind Valjoux 7734 movement.

Lip had approached Tallon in 1973 to give his vision to a few watches. The

---

YEAR OF RELEASE 1975

MOVEMENT Valjoux 7734 mechanical winding

RELATIVE VALUE ★ ★ ★

NOTABLE FEATURE Roger Tallon's ball pushers are colour-coded

manufacturer perhaps did not expect him to come up with something quite as radical, or as iconic, as the Mach 2000 collection – watches that look part modernist minimalism, part child's toy. This was true of even the most sedate of the various Tallon watches, such as 1975's non-chronograph Mach 2000 ref. 43765 model (with Duromat 7525 automatic movement). In many respects the designs were way ahead of their time – in, for example, their blacked-out cases, which would not become more commonplace in watch design for another decade. Indeed, the Lip Mach 2000 watches could be described as the first 'design' watches - with an aesthetic appealing to tastemakers and lovers of the very contemporary before it aimed to attract watch fans.

YEAR OF RELEASE 1975

MOVEMENT Duromat INT7525 movement by Durowe

RELATIVE VALUE ★ ★ ★

NOTABLE FEATURE The 'blacked-out' case was widely imitated

LIP SECTEUR

The Lip Secteur double retrograde jump hour was available in three case shapes, round (ref. 43651), rectangular (ref. 43647) and square (ref. 43649), though none of these watches was round, rectangular or square exactly either, each case having its own degree of asymmetry, its own assemblage of surface curves and angles.

Each of these models was powered by a Wester German-made Duromat INT7525 automatic movement made by the ETA-owned Durowe. This included a custom-built module – later also used in the Wittnauer Futurama, a watch distinctly similar to the rectangular Lip Secteur – that allowed the hands to be off-centre and to 'fly back' at the end of their run rather than rotate in circles (as with a standard analogue display).

This gave each of the watches its own take on a groundbreaking linear, fuel-gauge style of time display. It certainly looked interesting, but was not always the easiest to read; compare these with the relative legibility of the classically analogue 451R Duromat-powered models Lip was producing just a couple of years before.

Rather, these watches embodied the adventurousness in design that Lip would come to in later years. They certainly would not have been ideal for the timekeeping duties that Lip had undertaken for the Tour de France in 1959. Nor would such good-humoured inventiveness have worked for the armaments the company built for NATO and France, and which provided some 30 per cent of the firm's revenues by the middle of that decade – funds that allowed founder Fred Lipmann (later Fred Lip) to re-establish the name he had had in French watches before the war.

YEAR OF RELEASE 1972

MOVEMENT Ref. 43651, ref. 43647 and ref. 43649 Duromat INT7525 automatic by Durowe

RELATIVE VALUE ★ ★ ★

NOTABLE FEATURE Distinctive is the linear 'fuel gauge'-style display

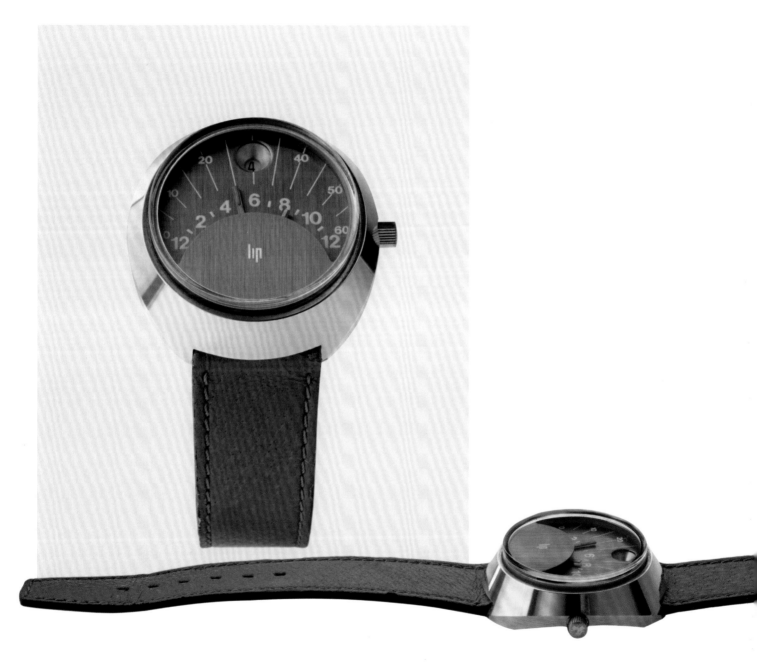

LIP SECTEUR

LIP SKIPPER

Not all of the watches designed for Lip by the group of seven industrial designers it hired in 1971 were as obviously radical as those designed by Roger Tallon or Isabelle Hebey. This Skipper model, with a mechanical in-house T-13 movement, was the more understated work of architect and designer Marc Held, who, given the relatively tame tank case, offered the simple yet certainly intriguing update of writing the name of each number on the dial out in full (an idea that was frequently lifted over the following decades).

Held's design typically leaned towards the restrained and thoughtful. He was the founder of the influential Paris-based design firm Archiform in 1960: his best-known designs included a rocking lounge chair for Knoll – designed with a curved base that could rock and swivel, all the better to ease the muscular strain that can come with sitting for too long – as well as beds, desks and chairs in fibreglass, tableware for Coquet porcelain, and a number of concept cars for Renault (one of which was the foundation of the category-defining Espace). Held would also succeed Marcel Breuer as effective designer-in-residence for IBM, designing most of the computer giant's major buildings in France.

YEAR OF RELEASE 1975

MOVEMENT Mechanical LIP T-13

RELATIVE VALUE ★ ★ ★

NOTABLE FEATURE Each hour number is written in full on the dial

LORD NELSON MYSTERY
+ LORD NELSON JUMP HOUR

Most watch design aims in various ways – through legibility, luminescence and so on – to make the reading of the time easier rather than more difficult. Not this watch from Lord Nelson. The piece (opposite), with a Swiss manual-winding Michael Berger Watch Company movement, requires the wearer to slide back a panel to reveal a mechanical jump hour digital display. Closed, the watch has the facelessness of a piece of jewelry; indeed, perhaps it encourages a healthier relationship to the measurement of passing time, making checking it more deliberate than merely reflexive. The watch returns an element of ceremony to checking the time, akin to that lost when (especially hunter-style) pocket watches fell from favour, or – for those who argue that the watch is redundant – when the pulling of a cellphone from one's pocket became ubiquitous.

Lord Nelson developed quite a reputation for bold designs. Less complicated than the Mystery watch perhaps, but no less striking, is this mechanical jump hour model (left) with a manual-winding movement, also from the 1970s. The piece perhaps inspired the case shape of the UR-202 model from avant-garde Swiss maker Urwerk, released more than three decades later in 2008.

YEAR OF RELEASE 1970s

MOVEMENT Swiss manual-winding mechanical

RELATIVE VALUE ★ ★ ★

NOTABLE FEATURE A panel must be slid open to reveal the time display (opposite)

LORD NELSON TRAPEZOID

Modern watch brands such as Bell & Ross and Panerai have caught the imagination by making outsized watches. But this is not a new idea. Dial time back to 1967 and Lord Nelson produced this manual-winding monster – some 50 mm (2 in.) long, with its heavy brushed plated-steel case all sharp angles, not least the one that slopes the dial towards the wearer, driver's-watch style.

Watches of unusually large dimensions used to be so as a question of functionality – as in early pilot's watches, in which case larger simply meant more legible – and occasionally for artistic expression, with more case and dial meaning more room for decoration. In some instances a larger size also meant that the movement need not be as compact (which is to say as advanced and expensive) as it might otherwise be; or, conversely, provided more room for more complications.

But once small-scale movement design became the mass-production standard – indeed, miniaturization has become synonymous with technological progress – case size became a matter of fashion, or of expressing status or even maybe masculinity. And it was certainly something of a trend in the late 1960s, with cases expanding well beyond the historical size for a man's watch of around 37 mm (1½ in.).

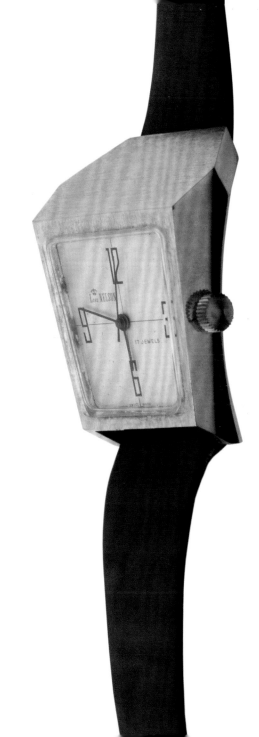

YEAR OF RELEASE 1967

MOVEMENT Swiss manual winding

RELATIVE VALUE ★ ★ ★

NOTABLE FEATURE One of the biggest watches of the era

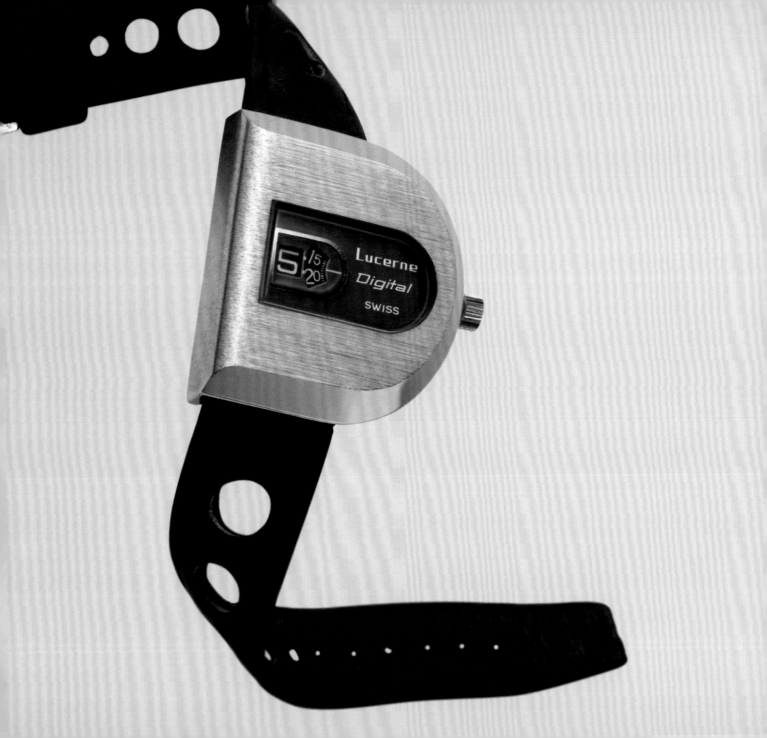

LUCERNE 'D' JUMP HOUR

One of a number of distinctive-looking watches produced during the 1960s under generic 'made in Switzerland' brands, this mechanical jump hour watch underscores the prestige of its country of origin by using the name of a Swiss city.

YEAR OF RELEASE 1960s

MOVEMENT Swiss manual winding

RELATIVE VALUE ★ ★ ★

NOTABLE FEATURE Takes its name from the Swiss city

MIDO CUSHION AUTOMATIC

Mido was given a Spanish name – it is from 'yo mido', or 'I measure' – but was founded by Georges Schaeren in Switzerland in 1917. For much of its early history the brand produced watches in the shape of radiator grills for automotive brands including Ford, Buick, Bugatti and Fiat, but by the 1930s it was introducing technical advances such as its 'Aquadura' system. This was a cork crown-sealing system that gave some of its models a degree of watertightness, a rare thing in watches before Rolex devised its Oyster, and an attribute it would later enhance by introducing the first monocoque watch case. During the 1930s Mido also produced one of the first anti-magnetic, water-resistant automatic watches, the Multifort, its best-seller for two decades, though perhaps its most unusual watch was not released until 1996: the Bodyguard, with an in-built 100-decibel security alarm. Comparatively, this cushion-cased Mido is less about function and more about style, being unusual for its vertical day/date display and the striated indices, with matching hands.

YEAR OF RELEASE 1960s

MOVEMENT Swiss automatic

RELATIVE VALUE ★ ★ ★

NOTABLE FEATURE Striated indices would become more commonplace in the 1970s

YEAR OF RELEASE 1967

MOVEMENT In-house manual-winding 17-jewel Cattin C66

RELATIVE VALUE ★ ★ ★

NOTABLE FEATURE Includes both magnetic compass and thermometer display

MORTIMA MAYERLING THERMO-COMPASS 'SURVIVOR' WATCH

Many might associate the idea of making this wrist-worn device called a watch do more than just measure time with the popularity of Casio's calculator watches during the 1980s, building on the possibilities afforded by liquid-crystal displays and the miniaturization of processing power. But the idea of the watch as a gadget – see James Bond for further inspiration – had occurred to watch designers ahead of the age of mass consumer electronics.

Take this Mortima Mayerling Thermo-Compass 'Survivor' watch, for example – so dubbed because this outsized piece might help you in a sticky situation. It is powered by a manual-winding 17-jewel Cattin C66 in-house movement, named after Césaire Emile Cattin, the founder of the Mortima brand's manufacturer Cattin & Cie in 1929. Cattin so changed the fortunes of Morteau, the town in eastern France in which he was based, that there is a road there named after him, alongside others named for the rather better known Charles de Gaulle, Albert Camus and Louis Pasteur.

But this watch also comes with a magnetic compass and a thermometer display, made possible by the use of a steel bi-metal spiral that expands and contracts precisely with temperature change (and a similar idea to that which would later be used in the power cut-off mechanism inside automatic kettles). Small wonder that, from the 1950s onwards, Mortima developed a reputation for creating somewhat outrageous watches.

Indeed, this piece, first launched by the watchmaker soon after the creation of the Mortima brand, is not all functional. The face of the chrome-plated steel case is heavily machine-etched with a highly textural pattern. Perhaps this doubled as a nail file. A version with a case in gold was also produced.

MOVADO ZENITH WOOD DIAL

Over the twentieth century watchmakers frequently produced timepieces under multiple brand names, and makers often produced watches for other brands or for retailers, but rarely were watches produced under a co-branding between two makers, as with this Movado/Zenith wood-dialled watch (with Zenith calibre 2572PC automatic movement).

Yes, from 1969 both brands belonged to the same holding company. But, all the same, this was an unusual arrangement. Indeed, the two brands collaborated on and off for fifteen years, between 1969 and 1984, basically with a deal that allowed them to use certain of each other's in-house movement designs. Zenith got to use, primarily, Movado calibres 405 and 408, while Movado was free to use Zenith's calibre 3019, better known as El Primero.

The ties went deeper on occasion too. Thanks to Zenith Electronics being an established brand in the USA, Zenith the watchmaker needed another brand name under which to sell its El Primero watches there, and Movado (in part) provided the answer. It would not be until 1984, when the Movado brand was acquired from the holding company, that the Zenith name started to appear alone on watches again.

YEAR OF RELEASE 1970

MOVEMENT In-house Zenith calibre 2572PC automatic

RELATIVE VALUE ★★★

NOTABLE FEATURE A rare example of a watch carrying two brand names

NOBREZA
MECHANICAL DIGITAL

From a design perspective, a mechanical digital watch such as this one from Nobreza – a Portuguese brand of the 1960s and 1970s, which used Swiss mechanical movements – presents something of a challenge if you also want to display the date. There is a risk of there simply too many digits on display at once. This piece successfully solves the issue by graphically separating time readout and date window, each highlighted by its own battleship-grey backing. Add in the cobra-head case shape and starburst-style brushed-steel finish and it is a stand-out piece from a largely forgotten maker.

YEAR OF RELEASE 1972

MOVEMENT Swiss automatic

RELATIVE VALUE ★ ★ ★

NOTABLE FEATURE 'Cobra' case shape with starburst brushed-steel finish

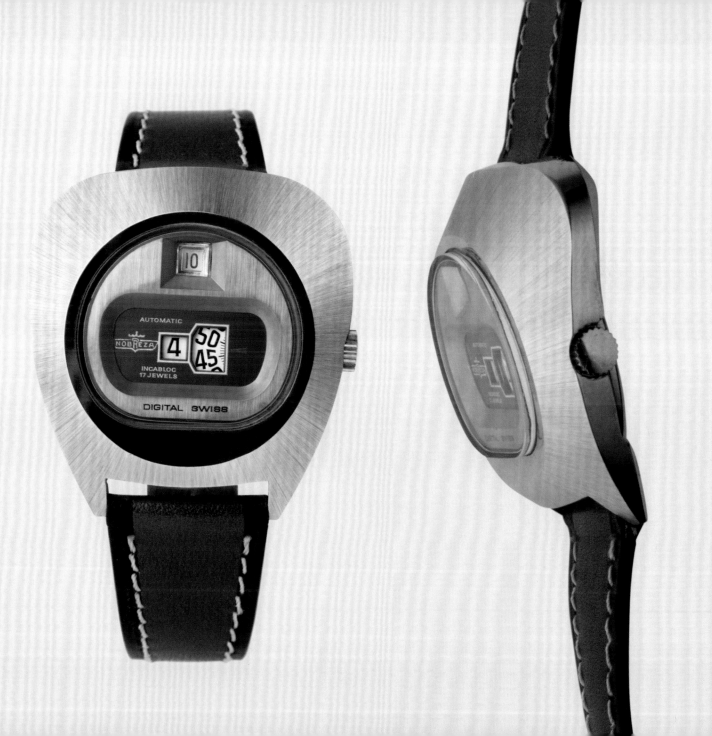

OMAX DRIVER'S WATCH

Omax, founded in Geneva in 1946, is probably best known as one of the makers of André Le Marquand's Spaceman watch design in 1974, but the company had by then an established history of making bold, even outrageous, designs both under its own name and for other brands, such that it was one of the Switzerland's biggest watch exporters of the 1970s. This driver's watch, with mechanical winding movement, is a case in point, with its asymmetric shape and, more obviously, its being made from a textured, rubberized plastic case in a vibrant shade close to International Klein Blue, developed by the artist Yves Klein ten years previously.

YEAR OF RELEASE 1970

MOVEMENT Swiss mechanical winding

RELATIVE VALUE ★ ★ ★

NOTABLE FEATURE Rubberized plastic allows for use of a bold case colour

OMEGA DRIVER'S WATCH

The driver's watch has come to be conflated with the chronograph – useful for timing laps, yet hardly designed with the physical act of driving in mind. Celebrity has underscored this more simple definition – for example, Steve McQueen's association with Tag Heuer's Monaco, Paul Newman with Rolex's Daytona model – as have countless collaborations between watch and car brands, exchanging materials and design details but perhaps above all marketing might. Rolex founder Hans Wilsdorf recognized this when, in 1947, he persuaded Malcolm Campbell to wear an Oyster while setting a new land speed record.

Since then watches have been made, for example, as one-offs to celebrate one-off cars, or with devices that allow the wearer to unlock their car door. To cite just a few such collaborations, IWC has partnered with Mercedes, Tissot with Alpine, Richard Mille with McLaren, Oris with Audi, Baume & Mercier with Shelby Cobra, and Bremont with Jaguar.

Yet few such resulting watches consider the specific needs of the driver. While the first recorded motor race, the Paris-Rouen – organized as a publicity stunt for *Le Petit Journal* – took place in 1894, it was not actually until the 1930s that drivers started to wear watches as a means of monitoring their performance. And, while a lap-timing complication has long been useful, it was perhaps only then that the benefits of a more considered, more ergonomic design became apparent.

Watchmakers have experimented with such ideas as turning a flat dial around its centre-point to a certain angle to facilitate viewing; moving the crown to a position so that it does not dig into the wrist when the hand is pushed back; and, of course, giving the time readout a side view – angling the time display towards the driver, so they do not have to take their arm off the steering wheel to see it. But this Omega De Ville, in solid 18k gold, takes that idea a step further. It not only keeps the dial simple for clarity's sake, but has a shaped case that allows the watch to be worn right on the side of the wrist.

YEAR OF RELEASE 1960s

MOVEMENT Unknown

RELATIVE VALUE ★ ★ ★
(solid gold case)

NOTABLE FEATURE The case back is shaped to allow wear on the side of the wrist

PALLAS QUARTZ + QUARTZ SEGTRONIC

YEAR OF RELEASE 1972

MOVEMENT Epsa Optel

RELATIVE VALUE ★ ★ ★

NOTABLE FEATURE The earliest liquid-crystal display (dynamic scattering LCD) – the 'missing link' between push-to-show LED and later traditional field-effect LCD

Some technological advances are bold efforts but perhaps destined to be stopgaps between the past and a more affordable, efficient and all too imminent future. Optel, Microma and Texas Instruments' development from 1967 of a display that moved on from LED – it did not require the pushing of a button to illuminate the time – without quite becoming the ubiquitous grey field-effect LCD that would later dominate the industry, is just such an advance.

Their system, used in both this colourfully dialled, unbranded watch (made by Pallas) and the Quartz Segtronic with crown, used what was known as dynamic scattering. Light that goes through the glass is absorbed by a mirrored back plate, both glass and back plate being attached to electrodes; when a charge is applied, molecular arrangement is disrupted, the light is scattered and this part of the display looks darker than the rest. Those segments that are not energized let the light pass through, reflect off the mirror and compose the digits. The problem? Aside from the fact that the digits showed in a low-contrast silver colour that was hard to see unless viewed square on, it was so power-hungry that new batteries would be required every eight months or so.

The system was so expensive that it had required six watch companies to collaborate to finance the research. So, once the investment had been made, perhaps the desire to press on regardless was strong; certainly the system's first, 7-volt iteration in 1972 would appear in watches produced by more than a dozen companies, including Waltham, Glycine, Zodiac and Sandoz. But only briefly. The research, it seems, had

primarily served as a spur to the parallel development of the field-effect LCD.

This was mainly the work of the International Liquid Crystal Display Company of Cleveland, Ohio, together with research bodies in the USA and Germany and Switzerland's Centre Electronique Horlogerie. Indeed, their efforts produced the first field-effect LCD watch – Gruen Watch Company's Teletime – in 1972. This, in brief, works by polarizing light passing through two filters into two perpendicular directions; a voltage is then applied across the glass plates, lining up the molecules to extinguish the light and leave part of the display dark, with the light part showing the digits. Since causing a change in the display takes only 0.3 seconds, it meant that seconds could be shown; what is more, the system only needed 1.5 volts to make it all work.

YEAR OF RELEASE 1973

MOVEMENT Epsa Optel quartz

RELATIVE VALUE ★ ★ ★

NOTABLE FEATURE First use of liquid crystal in a watch display

PAUL SMITH
DRIVER'S SIDEVIEW

Fashion-brand watches have tended towards affordability, quartz power and unchallenging design, placing more appeal in the brand name than the product. One exception to this was a collection of watches launched by British fashion designer Paul Smith, inspired by his own collection of vintage watches from the 1960s and 1970s.

YEAR OF RELEASE 1990s

MOVEMENT Citizen Quartz

RELATIVE VALUE ★ ★ ★

NOTABLE FEATURE A rare much later nod to the more unusual designs of the 1960s/70s

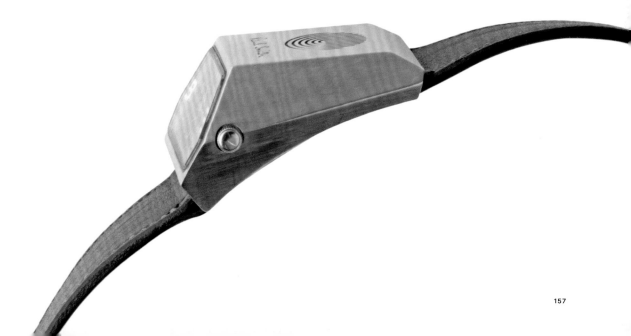

PIERRE BALMAIN

French fashion designer Pierre Balmain (1914–1982) was best known for making couture dresses, designing for such style icons as Marlene Dietrich and Katharine Hepburn, as well as creating outfits for the 1968 Winter Olympics and the airline TWA. But Balmain had also studied architecture at the Paris Ecole des Beaux-Arts during the 1930s, which perhaps helps to explain his brand's stylistically strong, asymmetric shapes for these watches from the early 1970s. The first moon landing had taken place just two years before the launch of the first of these watches, and while riding the aesthetic zeitgeist, they also mark a distinct break from the conservatism of much mainstream watch design of the time – a break perhaps made more easily by a fashion house than an established watch brand. Each Pierre Balmain watch came with a mechanical FE 233-60 or Lorsa F8 movement.

YEAR OF RELEASE 1970s

MOVEMENT Manual-winding mechanical FE 233-60 or Lorsa 8F

RELATIVE VALUE ★ ★ ★

NOTABLE FEATURE One of the earliest 'fashion brand' watch series

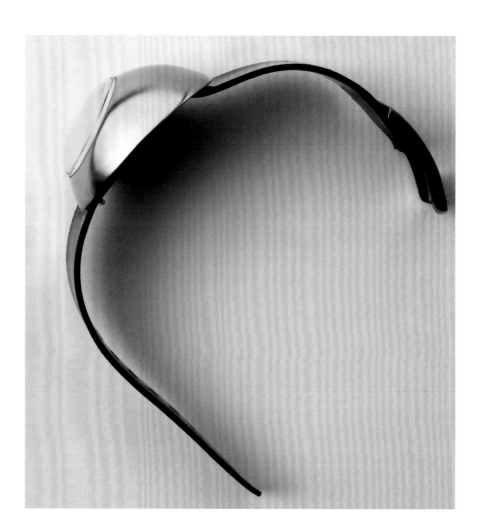

PIERRE CARDIN
ESPACE COLLECTION

Alongside other fashion designers such as Paco Rabanne and André Courrèges, Pierre Cardin (born 1922) took deep inspiration from the 1960s 'race for space' between the United States and the USSR. Courrèges was invited to Cape Canaveral by NASA; Rabanne designed the costumes for *Barbarella*; while Cardin – who even worked on a spacesuit prototype for the Americans – pioneered the use of silver-coloured fabrics, vinyl and outsized zips, and devised helmet-like hats with plastic visors. It was all part of his effort to design what he called clothes 'for a life that does not exist yet – the world of tomorrow'.

Cardin was also an enthusiastic licensor of his brand name (perhaps too enthusiastic, it would later prove) – hence the Espace collection of twenty-six watches. All featured a mechanical winding FE 68 movement signed by Jaeger-LeCoultre, but also offered some suitably futuristic ideas in watch design. The Chaffeur's Watch, for example, was a driver's watch with a case of such extreme curvature that it could be worn only on the side of the wrist. More strikingly, the collection was ahead of its time in offering interchangeable straps. The overall aesthetic was suitably 'out there' too: smoked crystal glass, layered discs, bifurcated cases, Lucite cubes, stripes, domes and nods to flying saucers all featured on various pieces. They came with some fantastic names too, chief among them surely the 'Espace Craterdome'.

YEAR OF RELEASE 1971

MOVEMENT Jaeger mechanical winding FE 68

RELATIVE VALUE ★ ★ ★

NOTABLE FEATURES A variety of futuristic case designs, minimalist displays and alternative materials

PIERRE CARDIN ESPACE COLLECTION

170 PIERRE CARDIN ESPACE COLLECTION

PIROFA 'BULLHEAD' CHRONOGRAPH

A short-lived manufacturer, founded in Nantes, France, by brothers Pierre and Robert Fontaine and operational only through the 1950s and 1960s, Pirofa nonetheless produced some striking pieces and in a huge diversity of styles, from ornate pendant fob watches to hardy diving watches, for which it would come to be best known. This 'bullhead' piece, so named for the pushers being mounted along the top line of the watch, was launched shortly before the company closed around 1970. This chronograph is powered by a Valjoux 7734 manual-winding movement, a reliable and sturdy mechanism widely used throughout the 1960s and 1970s, and which also allows the sub-dials to be set vertically.

YEAR OF RELEASE 1969

MOVEMENT Valjoux 7734

RELATIVE VALUE ★ ★ ★

NOTABLE FEATURE 'Bullhead' watches are named for mounting pushers along the case's top line

MECHANICAL VS QUARTZ

The watches in this book mark a definitive period in the history of the wristwatch: the transition, over the 1970s, from those powered mechanically, as watches had been for centuries, to those powered by battery. Perhaps Christmas Day in 1969 might be cited as the actual moment of change, when the Japanese company Seiko introduced the Astron, the first quartz wristwatch. It was a limited edition of a hundred pieces, hugely expensive – each costing roughly the same as a Toyota Corolla car – and, particularly because Seiko had given it a sweeping second hand, with a short battery life too.

Indeed, it was perhaps more a sign of what was to come rather than a revolution there and then, since Seiko would not release any more Astrons until 1971. But the effect on the Swiss industry – a few of whose makers were also launching quartz analogue watches around this time, but most of whom were too slow on the uptake of this new technology – was profound. Quartz movements were cheaper to produce, more self-sufficient (not requiring winding or regular servicing), and more accurate. They were a category killer.

And this was all the more so when quartz movements were teamed with a digital display, as the Hamilton Watch Company did in April 1972 with its Pulsar. This LED (light-emitting diode) piece was far from ideal – you still had to push a button to illuminate the time, and battery life was not great – but now here was a watch that looked like a product of tomorrow too. Within two years there were quartz models on the market, such as that by National Semiconductor, that were half the price of the Pulsar. By 1975 more than fifty companies were making LED digital watches in the US alone – none of them watch companies, all of them tech companies, including Hewlett-Packard, Intel, Hughes and Motorola.

By the end of the decade, makers of LED watches themselves had more or less dropped out of what was a dwindling market, to be replaced by makers of LCD (liquid-crystal display) technology. It was this display technology, creeping on to the market over the 1970s, that would really be set in opposition to the traditional mechanical movement.

OPPOSITE One of the first LCD watches, this Segtronic model marked a distinct improvement on the forerunning LED display, the latter both being power-hungry and requiring a button press for illumination.

ABOVE A traditional round, analogue display did not mean the watch itself needed to be a conservative design, as demonstrated by this 1973 Gubelin model, which was set into a steel bangle (pages 68–69).

ABOVE RIGHT The mechanical watch, then as now, came with the inconvenience of requiring regular and expensive servicing; in contrast, quartz models – while needing an occasional battery change – appeared all the more modern for being comparatively self-sufficient and contained.

RIGHT Quartz movements, while long associated with digital display watches, soon found themselves powering what, from the exterior, looked to be traditional analogue styles, like this 1969 Seiko, modelled closely on Seiko's pioneering Astron.

Often attributed to Japanese manufacturers, the quartz movement – such as this 1969 Beta 21 prototype – was in fact a collaborative effort between Japanese, American and Swiss watchmakers, even if the Japanese would most enthusiastically embrace this new technology.

Seiko – leader among the Japanese makers of quartz LCD technology – was back.

And in a big way. Despite demands for it to make LED watches, Seiko backed electronic movements and LCD to the hilt – complete with high-volume robotic production, also anathema to the quaint image then still favoured by Swiss manufacturers of its watches being made by hand by wizened craftsmen in Alpine chalets. By 1977, in sales terms Seiko was the world's biggest watch company, a product in part of the company's smartly putting its quartz movements into both digital and analogue models – the former largely overlooked by Swiss makers, the latter by American makers.

The Swiss, in particular, found responding to this existential threat – dubbed the 'quartz crisis' – almost impossible, largely because the industry was so fragmented, with each maker typically dependent on parts from multiple specialist sources. Swiss watch production almost halved in the ten or so years up to the mid-1980s. That is in large part why so many of the watch brands featured in this book disappeared, sadly taking their distinctive design sensibilities with them.

Saving the Swiss industry would require serious restructuring. And that is what happened, with Ernst Thomke streamlining multiple brands and subsidiaries under the giant watchmaker ASUAG into one new company by the name of ETA. ETA pushed into quartz movement production and showed it could take on Seiko – ETA's quartz Delerium model would, at 0.98 mm (²/₅₀ in.) deep, be the thinnest watch made – albeit with a series of bailouts from the banks. This is where one Nicolas G. Hayek came in, at the banks' behest. He proposed the merger of ETA with the other giant Swiss watchmaker of the day, SSIH, leaving this enlarged ETA to focus on movement production (it still dominates movement manufacturing today) while watch brands focused on designing and selling their wares, buying in their movements from ETA. Latterly brands have become more focused in developing their own in-house movements, both for credibility's sake, but in no small part also so as not to be beholden to ETA.

Thomke also had another big plan: to make a cheap quartz analogue watch with the clever idea of doing away with the movement's base plate and fixing the parts directly to the case back. This would save costs, as would making the case out of plastic. Hayek secured the funding for the idea – and the Swatch watch was born. And it was the phenomenal worldwide success of the Swatch that, in effect, saved the Swiss industry. It saved mechanical watches too, albeit as a side project. While quartz movements continued to account for the vast majority of watch movements produced globally, the Swiss, back in profit, could revisit what they knew best, mechanical watch movements, for the niche audience that still wanted them. This time the mechanical movement would be more for the watch aficionado than, as it had once been, for the everyday user.

By the early 1990s mechanical watches had started to gain a new credibility, in line with a growing appreciation for all sorts of craftmaking, and for all sorts of analogue technologies, despite their being outmoded in terms of sheer functionality, and their considerably greater expense. There are good reasons for this: the mechanical watch's durability – it has an heirloom quality to it, in contrast to quartz's seeming disposability; it is an expression of history and heritage – perhaps even of a simpler, pre-digital age; it often embodies a more attractive aesthetic than your typical quartz watch, given the latter's use generally of cheaper materials; and, of course, there is a romance to a mechanical watch precisely because, even as mechanical movements become more sophisticated, it kicks against the march of progress.

Things could have turned out very differently, with this book then celebrating the final years of the mechanical watch, before the idea was consigned once and for all to the museums. As it is, it celebrates more a moment in time.

Quartz watch movements are often associated with miniaturization. As the technology became ever smaller it allowed for ever more compact watch designs; but not all embraced this move, as this hefty, solid steel Bulova Accuquartz shows (pages 26–27).

MECHANICAL VS QUARTZ 177

PRISMA GT VALJOUX CHRONOGRAPH

Some watches are rarer than others purely by dint of the smallest detail. This is a 1970 Mondia GT, with a Swiss Valjoux 7734 mechanical movement, which is even signed Mondia. Only it is not a Mondia. The same watch much less often carried the external Prisma name, a brand specializing in affordable watches established in the Netherlands in 1948, and once promoted by the international footballer Johan Cruyff. Either way, the busy dial on this watch – a mix of track graphics, tachymeter and unusual asymmetric sub-dials with neon orange hands – makes it a stand-out piece of the decade.

YEAR OF RELEASE 1970

MOVEMENT Valjoux 7734 mechanical

RELATIVE VALUE ★ ★ ★

NOTABLE FEATURE The same watch was also available signed by Mondia

RACING HELMET VERTICAL DRUM DISPLAY

Not everyone wants a classic watch. Indeed, as with any other accessory, the field is open for a touch of novelty. Enter this mechanical digital jump hour watch, with a Swiss movement that, unusually, scrolls the time display vertically on a drum. The bulky case, obviously, is designed to mimic a racing helmet, so although it provides an angled view of the readout for any driver, it is unlikely to fit under the sleeve of their overalls.

YEAR OF RELEASE 1970s

MOVEMENT Unknown manual-winding mechanical

RELATIVE VALUE ★ ★ ★

NOTABLE FEATURE Time display is scrolled vertically using a drum system

RADO NCC 404

Rado's NCC, or New Construction Concept, range of eight watches was made between 1970 and 1978 and offered among the most distinctive of designs from a brand that became better known for its almost minimalistic reserve and, latterly, its pioneering use of ceramic. From an aesthetic standpoint, the likes of this 404 model – which had an ETA 2789 automatic movement – featured such details as an offset crown, an early use of Rado's revolving anchor logo (perhaps better known as a feature of its Captain Cook watches) and a 'starry night' copper-flecked dial. The dials, in fact, were what made this run of models visually most distinctive. Other NCC versions featured, for example, a dial in mottled lapis lazuli.

As for the new concept, this was the movement and stem's being encased in a snug-fitting elastomeric gasket, created with a view to improving water- and shock-proofing within the watch, rather than seeking to make the entire case waterproof, as had been standard until then. With NCC watches the front and back of the cases were simply fitted together using four screws, yet the rubber envelope inside ensured they could achieve an impressive 22 atm depth rating. The last of the NCC watches, the 505, added the advance of using a tungsten carbide case, an idea adapted to produce Rado's Diastar 515.

YEAR OF RELEASE 1972

MOVEMENT Swiss ETA 2789 automatic

RELATIVE VALUE ★ ★ ★

NOTABLE FEATURE Water-proofing was achieved by use of a rubber bladder

RECORD (LONGINES) PLASTIC AUTOMATIC

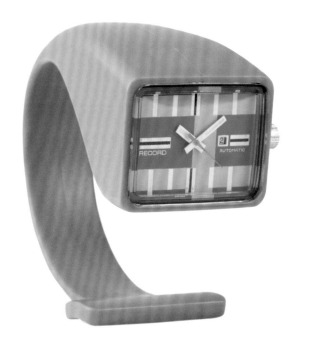
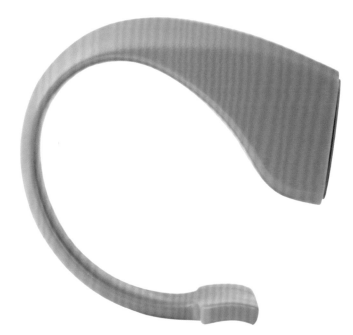

Watches of quality are typically associated with the use of suitably high-end materials, some appreciated for their inherent value, such as gold and platinum, others for their advanced performance properties, for example ceramic and carbon fibre. The use of plastic is more readily – but not always justly – associated with disposability. Or with fashion.

Such was the radical thinking when Nicolas Hayek came up with the Swatch watch in 1983: here were watches that, thanks to their use of plastic and mass-manufactured quartz movements, could be affordable and even collectible. A Swatch watch would be an alternative to your 'proper' watch. Indeed, Swatch takes its name from 'second watch'. But the use of plastics also changed the way a watch might look: plastics not only allowed the use of an endless spectrum of colours, but also could be moulded into bold new shapes.

While Swatch had both excellent reading of lifestyle aspirations and marketing on its side, this was not a new idea. Other companies had dabbled with plastic watches, sometimes decades before, among them Longines, whose

YEAR OF RELEASE 1968

MOVEMENT Swiss automatic

RELATIVE VALUE ★ ★ ★

NOTABLE FEATURE Plastic allowed a watch collection to become a reality for many

Record brand included such visually strong and innovative pieces as this sideview ladies' watch. Indeed, in 1971 Tissot launched the first watch with a nearly all-plastic movement, the Astrolon 2250 – the first time it had been possible to precision-mould the tiny parts with tolerances of less than one hundredth of a millimetre, and in a way that greatly reduced both manufacture and assembly costs. It even did away with the need for lubrication, something of a holy grail of more conventional, metals-based movement design.

Attitudes to both the Record and the Tissot watches were not entirely positive – one newspaper billed the latter as the 'throwaway watch, a timepiece with plastic works so inexpensive that they can be discarded and replaced whenever they fail to function properly'. But these watches reflect how plastics have allowed a more free-wheeling experimentation in form that commercial considerations largely prohibit when using more lasting and traditional materials.

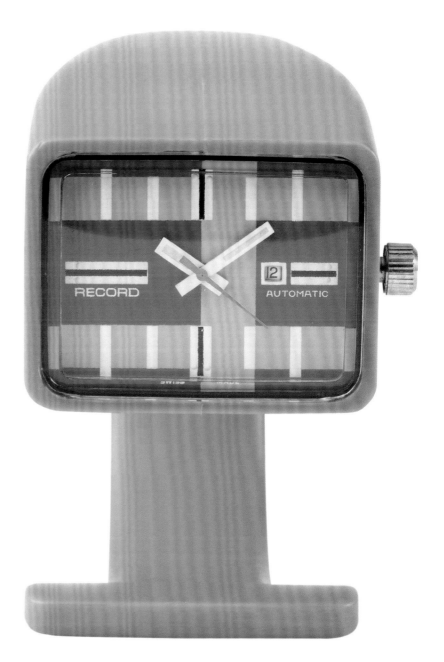

YEAR OF RELEASE 1970

MOVEMENT Manual-winding 1601

RELATIVE VALUE ★ ★ ★

NOTABLE FEATURE Designed by watch design legend Gerald Genta

ROLEX MIDAS CELLINI

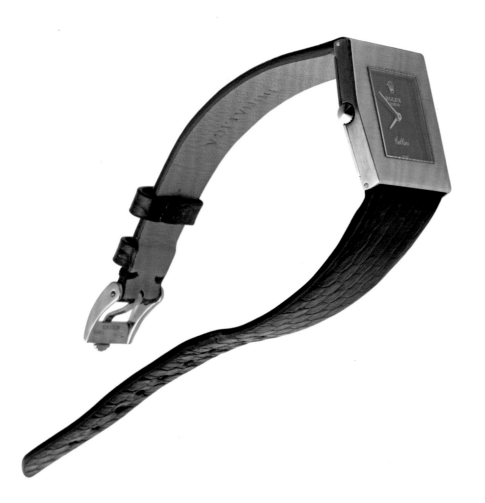

It is one thing to create an asymmetric case, another to extend the line of that asymmetry right along the strap all the way to the buckle too, as with this Midas Cellini watch with in-house manual-winding 1601 movement. The idea – a version of the limited-edition King Midas – was originally that of legendary watch designer Gerald Genta (1931–2011). The original, larger design, as its Greek myth-inspired name suggests, was carved from a single block of 18k yellow or white gold, and was designed to be worn on the right wrist, by left-handed people.

Charles Gerald Genta, to use his full name, had something of the golden touch himself. He may not personally have liked wearing a watch, but he designed classic after classic, including the Royal Oak for Audemars Piguet, the Ingenieur for IWC, the Constellation for Omega (as well as the case for its Seamaster), the 222 for Vacheron Constantin, the Pasha for Cartier, the coin watch for Corum, the Bvlgari Bvlgari for Bulgari and the Golden Eclipse and the Nautilus for Patek Philippe – the last watch sketched out over a five-minute lunch break during a Baselworld watch trade fair. And this is not to mention other designs for the likes of Chaumet, Hamilton, Breguet, Piaget, Rolex and Timex: so many designs that, as Genta once put it, 'I do not have recollection of all the watches that I've designed [and] I regret only not having designed the Rolex Oyster, because it represents the biggest success in watchmaking in terms of making a stylistic breakthrough'.

He also drove a breakthrough in the Swiss watch industry, getting it to appreciate that a designer who did not come up through the ranks of the watch industry could offer a fresh perspective on watch design. In an interview in 2009, Genta revealed his long diplomatic wranglings with the industry, given its problems with his being 'someone who is not an insider', as he put it. 'You know, it's very difficult to knock on the door of a prominent company and say "this is what I propose you do"! It's very pretentious. It's a delicate situation and you will never be recognized for your talent. I had to wait very patiently [for that].'

ROLEX OYSTER

When, in 1927, Mercedes Gleitze became the first woman to swim the English Channel, her success was somewhat marred when a hoaxer claimed to have made the swim in a faster time. So, just a fortnight later, she set out to swim the 35 kilometres (22 miles) or so again in what was dubbed the 'Vindication Swim'. It was already receiving considerable publicity when Hans Wilsdorf, founder of a watch company, realized the marketing potential of sponsoring her, and asked her to wear one of his new waterproof watches. Gleitze did not complete the crossing, needing to be pulled from the cold water 11 kilometres (7 miles) short of the shore. Nonetheless, a journalist from the *Times* noted that she was wearing a gold watch on a ribbon around her neck. It was still keeping perfect time. And a legend was born.

One month later, the Rolex Oyster Perpetual – akin to this later-date model, with a 1570 calibre automatic movement – was launched in the UK. Wilsdorf had founded the Rolex brand in London in 1908. With the creation by Cartier of what is recognized as the first wristwatch in 1904 – for the aviator Alberto Santos-Dumont – wristwatches began to grow in popularity as a replacement for pocket watches, and Wilsdorf set about creating a new, tough breed. In 1910 Rolex's Oyster received the Swiss Certificate of Chronometric Precision – a first for a wristwatch. And when Rolex bought the rights to the patent for a screw-down crown from Swiss watchmaker Perret & Perregaux, Wilsdorf's dream for the Oyster could finally be realized: in 1926 came the first dustproof, airtight and, most impressively, waterproof watch.

The endlessly emulated design was of such popularity that during World War II British prisoners of war could write to Rolex and order an Oyster, with the company taking each man's word as his bond and duly sending out the watch. Apparently it boosted morale, since taking payment on account implied a confidence that the Allies would win the war. Legend has it that one POW used his Oyster to time sentry movements and help conduct the famed Great Escape from the Stalag Luft III camp...

YEAR OF RELEASE 1960s

MOVEMENT Automatic mechanical calibre 1570

RELATIVE VALUE ★ ★ ★

NOTABLE FEATURE As worn on the first solo female swim across the English Channel

ROYCE 'MEXICO' WORLD CUP WATCH

Royce was, among the likes of Amphibian, Dorna, Eska, Budy, Leadership and the fantastically named Bonostar, just one of the many brands out of the Kocher, Sylvan & Cie Manufactory, founded by Sylvan Kocher in Selzach, Switzerland, in 1918. The company produced a diverse range of styles throughout its history, but always aimed to be at the stylistic cutting-edge, with 'les procédés de fabrication les plus modernes', as it put it in a 1972 advertisement. From strikingly minimalistic pieces in the 1950s through to more avant-garde case shapes during the 1970s, the little-known Royce – a brand created specifically for export to the North American market – was often ahead of its time.

And such was true of this huge, monobloc plastic-cased, space age watch, produced in celebration of the Mexico World Cup of 1970 and using the event's official font (designed by Lance Wyman, based on the one he created for the 1968 Mexico Olympics) on its dial. In keeping with the decade's exploration of less traditional materials in watchmaking, this Royce manual-wind watch was afforded a number of striking design details, from the unusual grey colour to the integrated strap and the asymmetric 'helmet'-shaped case. Note how this shape is underscored by the block indices, which increase in size towards the wider top of the dial.

YEAR OF RELEASE 1970

MOVEMENT Swiss manual winding

RELATIVE VALUE ★ ★ ★

NOTABLE FEATURE The integrated strap and asymmetric 'helmet' case

ROYCE 'MEXICO' WORLD CUP WATCH

SHEFFIELD OUTSIZED
+ SHEFFIELD JUMP HOUR

YEAR OF RELEASE 1970s

MOVEMENT Swiss manual winding

RELATIVE VALUE ★ ★ ★

NOTABLE FEATURE The 'Cyclops' model's teardrop-shaped case

Another example of the 1970s trend for oversized watches, this watch, with a Swiss manual-winding movement and a Corfam strap, comes up to a hefty 55 mm (2 in.) across. The use of chocolate brown for the strap, case inner and indices, and a mustard shade for the hands, seems particularly of the period, being a favourite for home decoration then too. Sheffield, an American brand established in the 1950s, had a reputation for producing affordable yet visually striking watches. A prime example might be the 'Cyclops' jump hour model on the next page, also with a Swiss manual-winding movement, in its gilt-plated teardrop-shaped case, with the hour and minute indicators each given their own display window (helpfully labelled 'hours' and 'minutes').

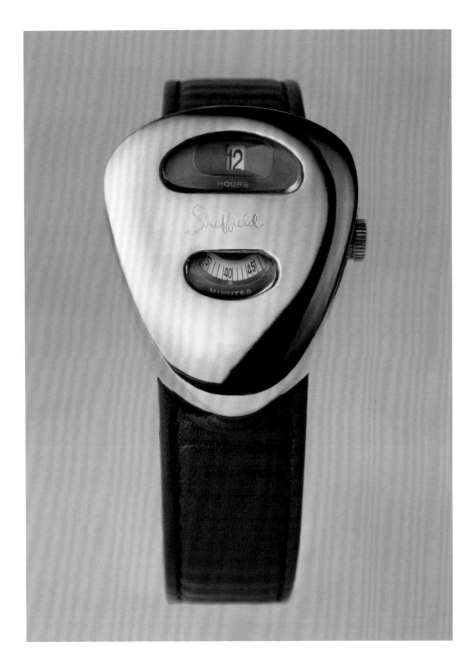

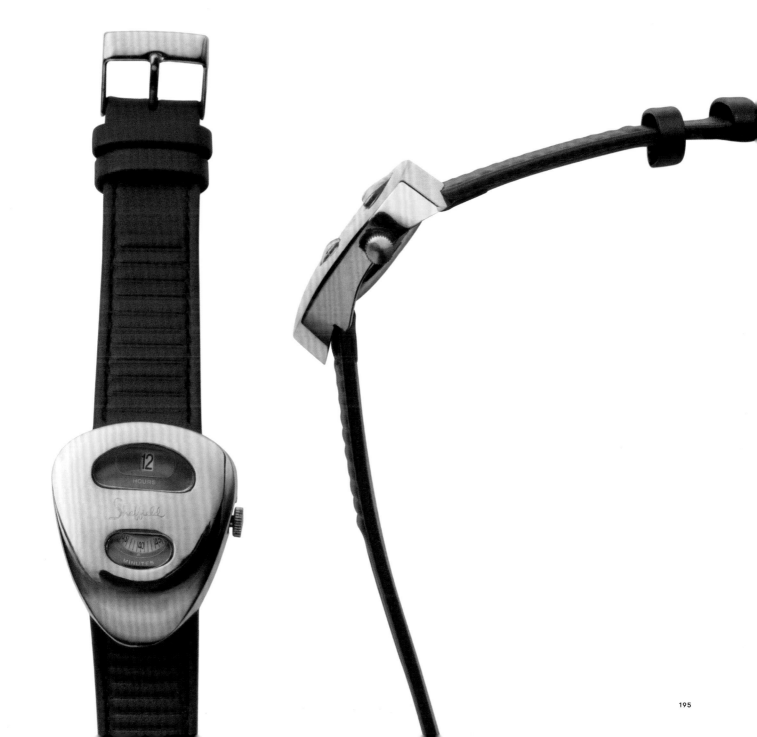

SICURA JUMP HOUR

Rather than show all time information in one window, this jump hour watch, with a BFG 158 Swiss automatic movement, makes a feature of the various elements by giving each its own window. To the left is the hour, in twenty-four-hour mode, marked 'H'; to the upper centre is the minutes display, in bright orange; to the right, in red, is the date, marked 'D'; and then there is an analogue second hand at the centre, albeit without any surrounding markers, so that it operates more as an indicator of passing time or that the watch is working.

The result is a mishmash of approaches from Sicura, a brand founded in Grenchen, Switzerland, by Théodore Sfaellos but really progressed by his engineer son-in-law Ernest Schneider, who took over in the early 1960s. By the mid-1970s he had built the company up to become a sizeable business, with four watch assembly factories.

Yet Schneider is better known for stepping in to save a troubled brand called Breitling, which was on the brink of closing in 1979. Schneider bought the Breitling trade name, and that of its most celebrated watch, the Navitimer, and set about reviving the brand, aided in no small part by the launch of its Chronomat, Aerospace and Emergency models. Sicura somewhat fell by the wayside. Schneider remained president of Breitling until he died in 2015.

YEAR OF RELEASE 1970s

MOVEMENT BFG 158 Swiss automatic

RELATIVE VALUE ★ ★ ★

NOTABLE FEATURE Each component of the time display is afforded its own window

SORNA MECHANICAL DIGITAL

At first glance this Sorna watch, with a Swiss manual-winding mechanical movement, looks to be rather conventional, its round case and that bold orange hand suggesting a standard analogue display. Look closer, however, and it is a jump hour display, with the hand for seconds display only, against unusual chequerboard-effect markers. Sorna was founded in the 1950s in Grenchen, Switzerland, and also produced watches under its Sornadat, Sornana and Sorina brands; but, at least in its original incarnation, it went out of business in 1994.

YEAR OF RELEASE 1970s

MOVEMENT Swiss manual-winding mechanical

RELATIVE VALUE ★ ★ ★

NOTABLE FEATURE Distinctive chequerboard-effect markers

SPACEMAN WATCHES

At the height of the space race a futuristic bent informed the broad spectrum of design. And so it was with watches too. In the late 1960s, Claude Lebet, owner of the Catena watch brand, based in Bulle, Switzerland, wanted to produce a watch that would be his own nod to the race to the Sea of Tranquillity. He asked André Le Marquand, an artist/architect based in Neuchâtel, to produce some designs, and the Spaceman was the result. The watch was introduced in 1972, which would turn out to be the tail end of the moon missions, but it was, nevertheless, like nothing the industry had seen before.

The watch, which had a manual-wind ETA movement, came in a large oval case, with a suitably progressive triple-forked Corfam strap by DuPont, and a coned dome crystal half-covered by a fibreglass section that allowed the dial to be visible only to the wearer. It was, of course, intended to mimic the half-closed visor on an astronaut's helmet. On some models indices and second hand were finished in bright orange – a very 1970s choice.

Multiple variations of the Spaceman were created, including the square-cased, metal Audacieuse (Invincible) of 1974, designed to give the line a more upmarket appeal. Others came in black gunmetal or on an asymmetric leather strap or with an automatic movement, and were sold under other brand names, including Omax, Jupiter, Fortis, Tressa and Zeno. The many variations encouraged collecting. Le Marquand would himself go on to establish his own watch brand, with boldly shaped cases one of its signatures. Among its designs was the Halfmoon, a more restrained version of the original Spaceman.

YEAR OF RELEASE 1972

MOVEMENT ETA manual winding

RELATIVE VALUE ★ ★ ★

NOTABLE FEATURE Domed crystal glass half-covered by fibreglass section

SULTANA DOUBLE RETROGRADE

The case for this double retrograde automatic is unfussy and minimalistic, all the better to highlight the attractive display, with the crystal cut into a trapezoid shape that is mimicked in the dial markers. Just as striking are the unusual colours – two shades of sky blue and muted orange. It is not the most intuitive of arrangements for a speedy reading of the time, but visually the display is certainly distinctive.

The Sultana brand was created in 1955, the product of a watch manufactory established in La Chaux-de-Fonds, Switzerland, in 1937 by one Paul Gaston Schwarz. From the outset the manufactory output was especially strong in sales to southern Europe and the Middle East: it won the Grand Prix at the Thessaloniki International Fair in Greece in 1939, and produced watches with both Turkish and Arabic hour markers. It was also more than competent on the technical front. One of its first, pre-war models was a mechanical watch with a single movement that could simultaneously display two time zones. It was one of the first companies to use a small magnifying glass built into the crystal to allow a clearer view of the date window, an idea now closely associated with the Rolex Submariner and similar models.

YEAR OF RELEASE 1970s

MOVEMENT Swiss ETA 2784 movement with custom module

RELATIVE VALUE ★ ★ ★

NOTABLE FEATURE Trapezoid crystal and dial markers

SWANK WATCH

Established in 1897, the Attleboro Manufacturing Company was a maker of men's and women's accessories, launching its Swank brand in 1927 and forming Swank Products Inc. in 1936, though not enjoying its heyday until World War II was over and the company had decided to ride the Mad Men era and focus on items for men. Historically it produced the likes of small leather goods, tie-pins, cufflinks and other men's jewelry for such brands as Pierre Cardin, J. C. Penney and Sears Roebuck, but likewise also designed watches, and had them made by Swiss contract makers with Swank branding on the dials. This arrangement allowed it to make short order runs and respond fast to changing trends – hence this 'woody' model, with a mechanical manual-wind movement and solid wood case and dial.

YEAR OF RELEASE 1960s

MOVEMENT Mechanical manual winding

RELATIVE VALUE ★ ★ ★

NOTABLE FEATURE Both dial and case are in solid wood

TELL FLEURIER SWISSONIC

The Tell brand dates to 1895, just a year after the founding of the Fleurier Watch Co. behind it. The municipality of Fleurier, in the Val-de-Travers, had been a Swiss watchmaking centre since the 1730s, but its reputation expanded rapidly after the opening of the prestigious Fleurier Watchmaking School in 1851. Indeed, by 1930 there were some thirty watchmaking brands in the Val-de-Travers, a fact echoed in brands operational there today, including Parmigiani Fleurier, Vaucher Manufacture Fleurier and Bovet Fleurier. But having a long history does not necessitate traditional design – as this Tell watch, with an ESA 9158 electronic balance wheel movement, suggests.

YEAR OF RELEASE 1970s

MOVEMENT ETA - ESA 9158 electronic balance wheel

RELATIVE VALUE ★ ★ ★

NOTABLE FEATURE Graduated grey dial and raised indices

TEVIOT JUMP HOUR MECHANICAL DIGITAL

With the advent of quartz movements and the subsequent fashion for digital displays, many watch brands sought to compete by giving the same style of readout to their mechanical pieces, in a brave, if largely futile, attempt to keep pace with the times. Substituting hands for discs on top of a traditional movement – in this case the AS2083 – was simple enough, and resulted in some graphically bold jumping hour designs, not just from Teviot, but also from Sicura, Technos, Nelson, Sheffield, Dugena and Ravisa, among many small and now long defunct brands. A similar watch to the one pictured was produced by Teviot with the dial arrangement flipped and the date window topmost.

The idea of the mechanical digital watch was not a new one: the first known example was a pocket watch created by French watchmaker Blondeau in the 1830s for the French king, while such makers as Breguet and Le Roy similarly pursued this new idea. Mass-produced 'jump hour' pocket watches were popular in the 1890s; in 1881 IWC had taken a gamble when it took on an invention by an Austrian father and son team – both named Josef Pallweber – of an under-dial module that used digits painted onto discs, visible through windows in the watch face, with the discs 'jumping' every sixty seconds. IWC promoted its new, considerably more expensive pocket watch as 'the watch without hands'.

The first digital-display wristwatches are believed to have been produced by Cortebert and Audemars Piguet in the 1920s. Then, as in the 1970s, having to show only four digits allowed the digital display to take on a decidedly clean look (and, some argued, also made for a much easier way to read the time than the unintuitive analogue display). But while the first wave of mechanical digital mania was regarded as excitingly progressive, in the 1970s it perhaps smacked more of desperation. And the quartz digital tsunami simply washed over it.

YEAR OF RELEASE 1972

MOVEMENT Swiss AS2083 mechanical

RELATIVE VALUE ★ ★ ★

NOTABLE FEATURE One of a number of mechanical digitals created to stave off the 'quartz crisis'

TIMSHEL MECHANICAL DIGITAL

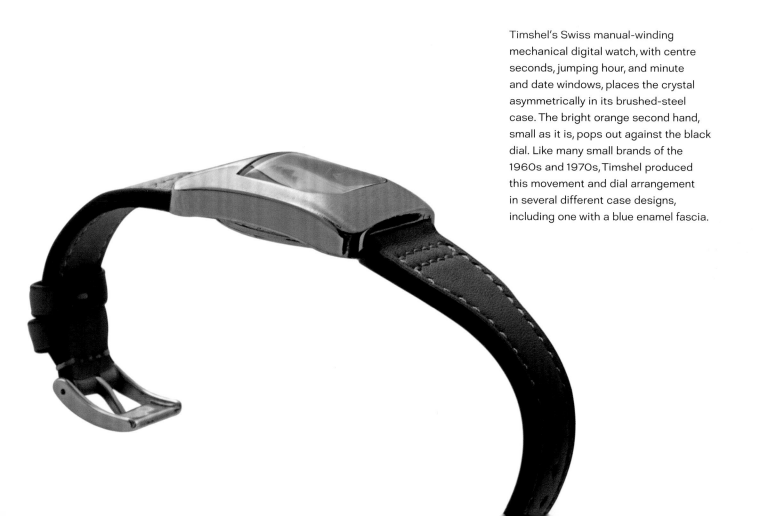

Timshel's Swiss manual-winding mechanical digital watch, with centre seconds, jumping hour, and minute and date windows, places the crystal asymmetrically in its brushed-steel case. The bright orange second hand, small as it is, pops out against the black dial. Like many small brands of the 1960s and 1970s, Timshel produced this movement and dial arrangement in several different case designs, including one with a blue enamel fascia.

YEAR OF RELEASE 1970s

MOVEMENT Manual-winding mechanical

RELATIVE VALUE ★ ★ ★

NOTABLE FEATURE The use of 'rescue' orange for the central second hand

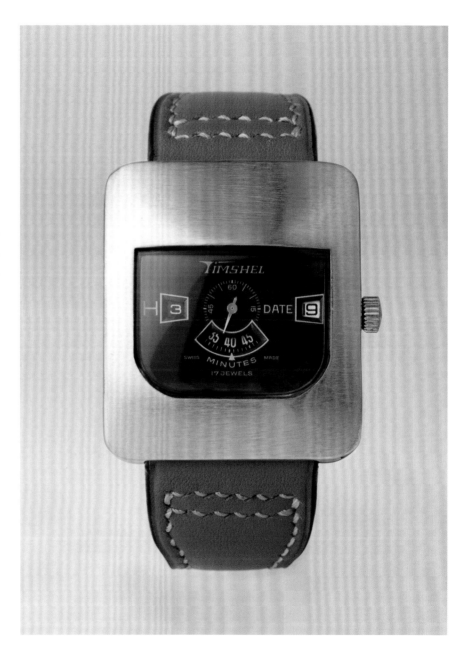

TIMSHEL MECHANICAL DIGITAL

TISSOT SYNTHETIC IDEA 2001

New materials, or at least those not commonly put to certain purposes, afford new thinking. The traditional Swiss watch industry long shunned even the latest plastics, for example, not simply because of their functional limitations but because of their association with cheapness. Tissot made a bold step towards changing that in 1971, when it launched its Tissot Synthetic Idea 2001, the first watch to come not just with a plastic case and strap, but with a movement almost entirely made out of plastic too (excluding core components such as the balance, barrel and mainspring). That choice of '2001' was probably no accident, suggesting the future in much the same way as Stanley Kubrick's film of three years earlier had.

It was the development of systems that allowed the milling of plastic parts to tolerances of within one

hundredth of a millimetre that made such a movement possible. But this was more than a gimmick. Using plastic components meant the movement could be built using fewer parts – just fifty-two – and without screws, further reducing the need for servicing; it could be built more quickly too, which reduced costs. It made for a naturally anti-magnetic watch and a lighter one – typically regarded as a property of good design, even if (often unnecessary) weight has long been marketed to be associated with quality. And this in-house Astrolub/Astrolon 2250 calibre even did away with the need for lubrication – something of a holy grail of more conventional, metals-based movement design. That was one patent Tissot would certainly file for.

Remarkably, Tissot had started work on this project in 1952. Those twenty years of work were well worth it in terms of moving watchmaking on, even if the Synthetic Idea 2001 proved to be a commercial flop. Tissot, it is said, failed to communicate the design's advantages over conventional watchmaking, while the public struggled to connect a material it associated with cheap children's toys with a personal product often worn to evoke prestige. Tissot sold some 15,000 pieces a year for four years before calling time on the idea. It was left with an unsold inventory of 500,000 watches. The watch was just too far ahead of its time – even if it did inspire the phenomenal success of Swatch some twelve years later.

YEAR OF RELEASE 1971

MOVEMENT Tissot in-house plastic

RELATIVE VALUE ★ ★ ★

NOTABLE FEATURE Tissot began work on this project in 1952

TITUS DOMED AUTOMATIC

An oddity among even the more extreme watch designs of the 1970s, this Swiss automatic watch from Titus has a steeply domed case, making it sit very proudly on the wrist. The dial is packed with detail, from the tubular indices at the quarter hour to the punched baton hands and the starburst minute ring. And then there is the industrial articulated bracelet. All the same, it is the case that steals the attention.

Solvil et Titus, the company comprised of the Solvil and Titus brands, was founded by watchmaker Paul Ditisheim in La Chaux-de-Fonds, Switzerland. He worked at his family firm, the watchmakers behind the Vulcain and Movado brands, until 1892, when he struck out alone. A specialist in chronometers, following his studies of the impact of magnetic fields and atmospheric pressure on their accuracy, with his designs Ditisheim would win the Royal Kew Observatory's prestigious world chronometric record in 1912.

In 1930 Ditisheim ceded control of Solvil et Titus to Swiss entrepreneur Paul Bernard Vogel, who successfully gave its two brands new direction, Solvil for haute horlogerie pieces, Titus for more mass-market watches. In 1968, with Titus, Vogel would go on to head the new Société des Gardes-Temps, a giant conglomerate of low-cost watchmakers and one of the biggest watchmaking companies in history.

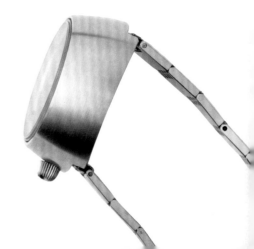

YEAR OF RELEASE 1970s

MOVEMENT Swiss manual winding

RELATIVE VALUE ★ ★ ★

NOTABLE FEATURE The indices and quarter-hour markers are cylindrical

DESIGN DETAILS

It might be argued that, while the brand giants of the mechanical watch world are somewhat limited in their aesthetic scope – being largely tied in to a particular look, one for which they are recognized but that also has sufficient commercial appeal to ensure that their profits outweigh their sizeable costs – smaller brands are free to take a more avant-garde approach. This was a defining characteristic of the countless makers of the 1960s and 1970s, before the 'quartz crisis' forced many to close. In a sense, with the advent of the Internet, the watch industry today is perhaps seeing the beginnings of a re-run of those same circumstances.

What made the 1960s and 1970s particularly expressive in watch design was, arguably, a consumer ready to buy a watch less for the brand name than for the way an individual watch looked. Advertising, pushing a select few makers over the many, was less abundant; broadly, watches were worn more for utility and personal expression than as status items, so the fact that few had heard of the maker of your watch hardly mattered. Meanwhile, the Swiss mechanical watch market was enjoying rude health, fruit of advances in manufacturing capability and expanding markets married with consumers wealthy enough to buy its product for pleasure rather than out of necessity.

In retrospect, the 1960s and 1970s – the decades that dominate the selection of watches in this book – brought together a number of design characteristics in watches that made many of them particularly striking. Of course, like the 1950s before it, the period produced watches that would go on to be considered classics, Zenith's El Primero, Patek Philippe's Nautilus and Audemars Piguet's Royal Oak among them. The period saw plenty of pedestrian watches too. But in between these were watches that struck a delicate line between being novelty and being playful.

Certainly one thing that unites the watches in this book is that, rather than being sufficiently archetypal or quintessential to be considered classic in the way those aforementioned watches are, they are all very much of their time. There is the

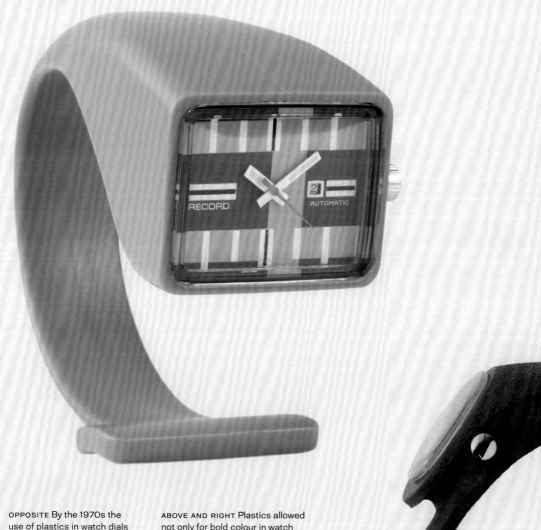
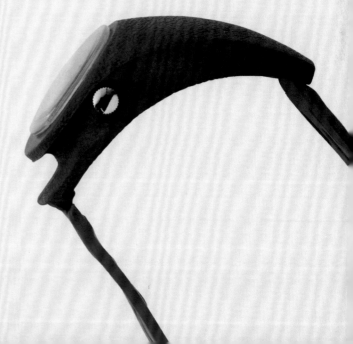

OPPOSITE By the 1970s the use of plastics in watch dials afforded a new variety of colours, as in the striking bulls-eye target of blues in this analogue watch from Webster (pages 236–37).

ABOVE AND RIGHT Plastics allowed not only for bold colour in watch design but for some distinctive forms too. This Record automatic piece from 1968 (pages 184–85) blurs the line between watch and bracelet, while the Omax one (right) plays with shape to make for a distinctive driver's piece (pages 148–49).

220 DESIGN DETAILS

use of colour, for example: bold hues, tone on tone shades, fade effects and the like are commonplace. Materials are no longer limited to stainless steel or precious metals – there are early experiments with plastic and with wood, for instance. Case shapes make a concerted break from the standard round towards bold oblongs, rounded squares, ovals and asymmetric shapes. There is a readiness to push the graphic potential – the sheer visual pleasure – in details as small and humble as, for example, the size, colour, form and positioning of indices.

This level of exuberance in watch design has rarely been seen since. Watches then were often fun-looking, a quality that since seems to be been lost in the pursuit of some high seriousness. But that sense of fun rarely detracted from their wearability or, indeed, their appeal – then or now.

If a mechanical watch was typically a watch for life, fashion designers inevitably encouraged the opposing idea of customers building a wardrobe of watches. To this end they pushed the expressive potential of materials and colours to make pieces of distinction such as this from Pierre Cardin's 1971 Espace collection (pages 164–171).

OPPOSITE By the early 1970s, movements were largely accurate and affordable – even unusual ones such as this Lip jump hour model (pages 114–15), seen from the front and in profile – and watchmakers turned to the expressive potential of materials, whether that be the hard and industrial quality of brushed stainless steel or, as with the Bulova Accutron of 1973 (pages 24–25), the warmer tones of dark wood.

TOSCA DRIVER'S WATCHES

Two iterations of essentially the same watch can result in a very different aesthetic feel, a point that watch brands such as Tosca were quick to exploit in order to get the most back on their development costs. The gilt-plated model with the burnish-effect dial and bold indices is more of a dress watch, while the model with the busier, wood-effect dial has a sportier tone. This latter watch also has a Swiss automatic movement. The other has a mechanical winding movement.

YEAR OF RELEASE 1969

MOVEMENT Swiss mechanical winding

RELATIVE VALUE ★ ★ ★

NOTABLE FEATURE The watch display is tilted up towards the wearer for better viewing while driving

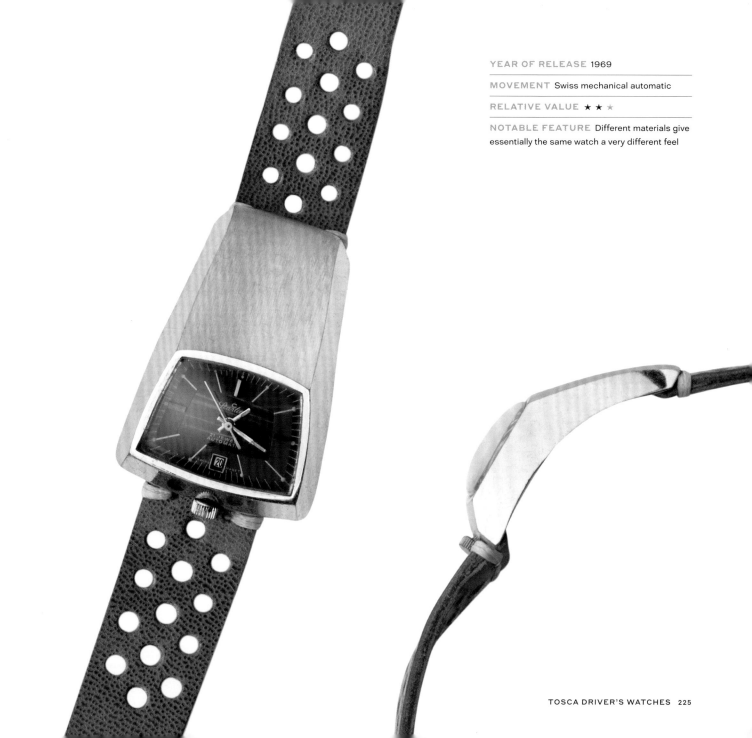

YEAR OF RELEASE 1969

MOVEMENT Swiss mechanical automatic

RELATIVE VALUE ★ ★ ☆

NOTABLE FEATURE Different materials give essentially the same watch a very different feel

UNION LADIES' WATCH

YEAR OF RELEASE 1973

MOVEMENT Swiss mechanical winding

RELATIVE VALUE ★ ★ ★

NOTABLE FEATURE Horizontally graduated shading over the dial

With most ladies' dress watches historically tending towards dainty circular cases, this piece from Union, with a Swiss mechanical winding movement, stands out for its rounded rectangular case – the shape emphasized by the subtle horizontally graduated colour of the dial. Although a number of brands have used the Union name, or variations of it, the watch was probably produced by Union SA Soleure, founded in 1895 in Selzach, in the Swiss canton of Solothurn (Soleure), by Max Studer. The company remained in production until the late 1970s but, struggling to survive the advent of quartz movements, was finally absorbed by Longines.

YEAR OF RELEASE 1960s

MOVEMENT Mechanical hand-wound

RELATIVE VALUE ★ ★ ★

NOTABLE FEATURE Unusual olive shade of dial

UNIVERSAL GENEVE TANK

Not all watch design of the 1960s and 1970s adopted the period's tropes of asymmetry, use of strong colour and new materials, atypical case shapes and so on. Some pieces avoided these yet still manage to remain characteristic of the times, such as this Universal Genève tank model, with hand-wound movement. It does so through a much more subtle approach, such as the rounded square of the case, or the olive shade of the fumé dial.

At the time this watch was released Universal Genève was still riding high on the profile it had achieved through the launch of the Gerald Genta-designed Polerouter in 1954. The company was in the midst of a fifteen-year run of multiple variations of the Polerouter, with its innovative anti-magnetic 'microtor' movement – ideal for flying over the poles, as international airliners had recently begun to do.

The brand had established a reputation as an innovator since its founding by Ulysse Perret and Numa-Emile Descombes in 1894, with Perret relocating the business to Geneva in 1919. Universal Genève became well respected as a maker of chronographs from the launch of its Compax model in 1936, and the 1950s and 1960s saw the brand collaborate with the French luxury-goods maker Hermès on a series of dress watches and Pour Hermès chronographs.

VULCAIN 'EYE' DIAL

Unconstrained by the essentially round nature of an analogue display, designers of mechanical digital watches during the 1970s felt free to make the glass on their watches in myriad shapes. And yet, more rarely, analogue pieces bucked expected form too, arguably at a cost to their utility. This Swiss automatic Vulcain watch sets its hands at the centre of an eye-shaped opening, drawing out the indices to follow the same shape. Intriguingly, while the date window is set clear of the hands, the day window is placed closer to the centre and so is covered at quarter to each hour – likely a compromise in the design of the movement.

Another of the watch brands that came out of the Ditisheim family of watchmakers, Vulcain was established in 1858. But it would become best known almost a century later, in 1947, for its launch of the first functional mechanical alarm movement – functional in the sense that its two-barrel construction allowed it to be the first that was loud enough to rouse any attention when sounding. It appeared in many iterations of its Cricket watch, dubbed 'the Watch for Presidents', owing to the number of US presidents who have worn it. While Rolex saw Edmund Hillary and Tenzing Norgay wear one of its watches on the first successful ascent of Everest in 1953, Uno Lacadelli and Achille Compagnoni both wore Crickets when they became the first to climb K2 the following year.

YEAR OF RELEASE 1970

MOVEMENT Swiss mechanical winding

RELATIVE VALUE ★ ★ ★

NOTABLE FEATURE Distinctive 'eye' dial shape and date display shading

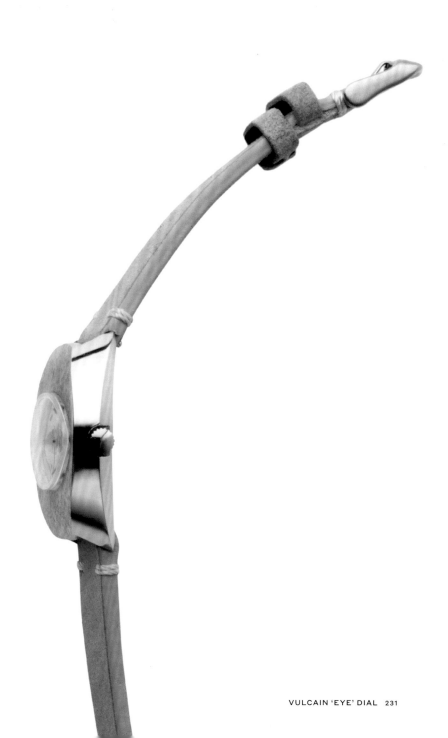

VULCAIN JUMP HOUR

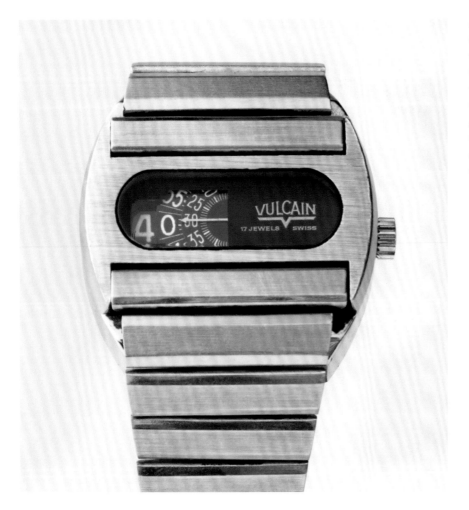

A mechanical digital jump hour watch from Vulcain: distinctive about it is the way the bracelet is integrated into the stepped case at its top, rather than into its side. This allows the case to be big enough to house the Swiss automatic movement yet to look, from the wearer's perspective, much more compact. Almost half of the display window is given over to branding.

YEAR OF RELEASE 1970s

MOVEMENT Swiss automatic

RELATIVE VALUE ★ ★ ★

NOTABLE FEATURE Bracelet integrated into the top of the case

WALTHAM JUMP HOUR CHRONOGRAPH

It is a rare thing to see an analogue chronograph combined with a digital time display. Yet here it is, in the form of this Waltham piece, complete with a calibre 1376 movement and in a yellow base metal can with 18k electroplated finish. In fact, this was possibly the first jump hour chrono and certainly the smallest chronograph of its day. Much the same watch was available in a steel case from Kelek SA, with a Tenor-Dorly movement and chronograph section built by Dubois-Dépraz, which supplied the same for Breitling's Chronomatic. It was one of the smallest chronographs on the market at the time.

This watch was made in greater numbers under the better-known Waltham brand, the American watchmaker established (then under another name) in Waltham, Massachusetts, in 1850, and noted for developing what would become known as the American System of Watchmaking – essentially manufacturing movements in such a way that they could be fully interchangeable between different watch models. Waltham garnered a reputation for making precision timepieces for the railways, aircraft manufacturers and Ford's more upscale cars, though exited the consumer watch market in 1958. A Swiss subsidiary had been founded in 1954, however, and continued to produce mechanical watches. Abraham Lincoln carried a key-wound Waltham pocket watch.

YEAR OF RELEASE 1974

MOVEMENT Kelek calibre 1376 automatic

RELATIVE VALUE ★ ★ ★

NOTABLE FEATURE The first jump hour chronograph

YEAR OF RELEASE 1970s

MOVEMENT Swiss Webster hand-wound mechanical

RELATIVE VALUE ★ ★ ★

NOTABLE FEATURE Bulls-eye effect in shades of blue

WEBSTER
HAND-WOUND

This Swiss mechanical winding watch from Webster is an exemplary instance of form and colour in harmony. The case shape may be traditional, but the dial creates a bulls-eye effect in shades of blue, with the hour and minute hands mounted on overlapping discs in toning colours. Webster started out as a maker of quality pocket watches but, by the mid-twentieth century, had fully turned its attention to the mass-manufacture of affordable watches. This model, however, suggests that accessible pricing need not exclude original design. Who was responsible for the design itself is unknown – designs under the Webster brand were also used in almost identical pieces under other long-lost brands such as Rego, Exactima, Rouan, Cimier and Ginebras.

WITTNAUER FUTURAMA

In its essential design the Wittnauer Futurama has much in common with Lip's Secteur model: its Swiss ETA 2784 movement includes a custom module that allows the flyback hour and minute hand to jump back to the top of the vertical display once its run from top to bottom is complete over its twelve-hour or sixty-minute cycle. Like Lip's model, the Futurama makes for one of the most striking and inventive pieces of the 1960s/1970s era – a last gasp of creativity before the rise of quartz movements somewhat curtailed such adventurousness.

Wittnauer's version differs subtly in the case, being asymmetric (as opposed to the Lip model siting its dial asymmetrically with a case that sits symmetrically on its strap) and with softer curves. Not every Wittnauer model was quite so futuristic; the Wittnauer 2000, launched in 1970, was a perpetual calendar that, for reasons unknown, only went up to the year 2000. By then perhaps, they imagined, everyone would carry a device in their pockets that told the time to absolute precision.

YEAR OF RELEASE 1969

MOVEMENT Swiss ETA 2784 movement with custom module

RELATIVE VALUE ★ ★ ★

NOTABLE FEATURE In effect an asymmetric version of Lip's Secteur model

WITTNAUER POLARA

At first sight one of the most curious things about this Longines-Wittnauer Polara, apart from that sleek case/bracelet integration, is its two-digit display. Inside was a Hughes Aircraft 2000 Series: an early LED module that meant the button had to be pressed once to display the hour, and then a second time to display the minutes. A year later and LED modules allowed the simultaneous display of hours and minutes with just one push.

Yet it is pertinent that this Wittnauer piece is a ladies' watch. Wittnauer, founded by J. Eugene Robert, had its roots as a watch import business based in Maiden Lane, the epicentre of New York's jewelry trade. In 1872 Robert was joined by his brother-in-law, Albert Wittnauer, who soon found a market for watches built by the company rather than just the Jaeger-LeCoultre, Vacheron Constantin and Longines pieces shipped in from Switzerland. The first Wittnauer-branded watches were launched in 1880.

It was a proper family firm: Albert was later joined by brothers Emile and Louis and, in 1916, after all three brothers had died, their sister Martha Wittnauer became the first CEO of a watchmaker and later the first woman elected to the Horological Society of America. Despite having no background in business, nor in watches, she drove the creation of Wittnauer's first truly pioneering model, the All-Proof, launched in 1926 and one of the first shockproof and waterproof anti-magnetic watches. It was a Wittnauer chronograph that, some forty years later, was selected alongside Rolex and Omega for final testing by NASA for astronauts to wear on the Gemini and Apollo missions.

YEAR OF RELEASE 1975

MOVEMENT Hughes Aircraft 200 Series LED Module A

RELATIVE VALUE ★ ★ ★

NOTABLE FEATURE Two presses of the pusher were required to show hour and then minutes

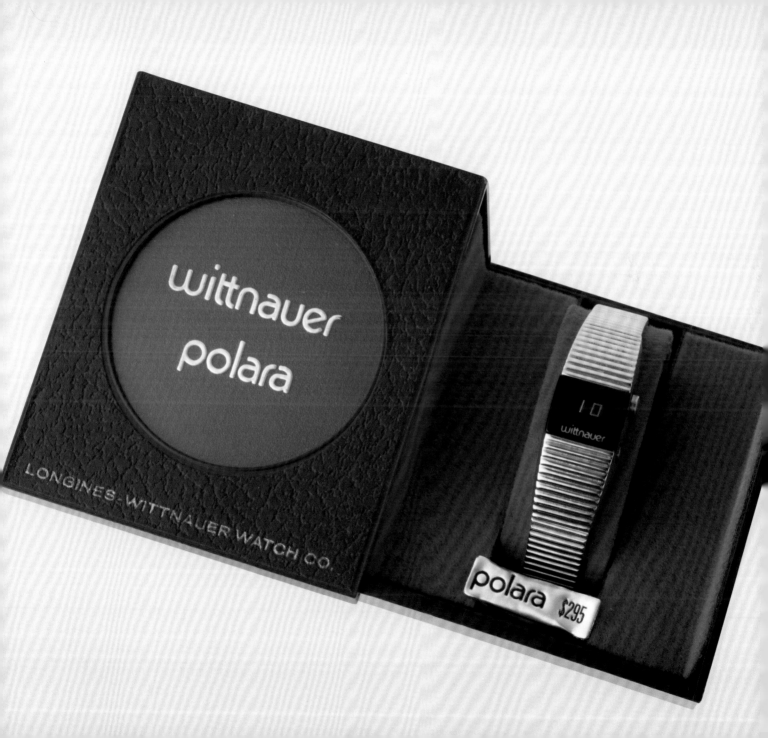

YEMA MECHANICAL DIGITAL

French brand Yema's men's and women's mechanical digital jump hour watches both have a dial comprising overlapping hour and minute discs, the time read through the red box and line indicators painted on the underside of the glass. Indeed, Yema watches were distinctive largely for their dial displays, as is also demonstrated by the jump hour model overleaf, with its three-window display for hours, minutes and date – cleverly and subtly sized in descending order of importance.

Yema, a company founded in Bescançon in 1948 by Henry Louis Belmont and the leading exporter of French watches in the mid-1960s, might take first prize for coming up with some of the most interesting watch names of the 1960s to 1980s, including the Yema Superman, Rallygraf (as worn by Formula One star Mario Andretti) and Spationaute, the first French watch to be worn in space.

YEAR OF RELEASE 1970

MOVEMENT Swiss mechanical winding

RELATIVE VALUE ★ ★ ★

NOTABLE FEATURE Red box and line indicators highlight where to read the time

YEAR OF RELEASE 1970

MOVEMENT Swiss mechanical winding

RELATIVE VALUE ★ ★ ★

NOTABLE FEATURE The displays for the hours, minutes and date subtly reduce in size according to importance

ZENITH TIME COMMAND FUTUR

Any brand that refers to a watch as the Futur really has to deliver something that looks like it is the shape of things to come, and Zenith surely pulled it off with this model. Part of the watchmaker's Time Command series of watches, the Futur combined both analogue display and digital seconds readout via the small portal bottom right of the watch face; it also had a fresh take on the period's love of asymmetry, with the top line of the plexi crystal seeming to curve over and disappear into the minimalistic bracelet. (Another version of the same watch, without this design detail, is much less arresting.)

This was, as Zenith announced in a press advertisement, 'the first quartz watch of its kind in the world... A new way to look at time'. The analogue display did not require the pushing of any buttons. But, the ad went on, 'when you need to see time with second-by-second precision, the digital display shows the passing seconds on command'. When setting the watch for a new time zone, the hands also automatically reset, while a memory circuit counts the passing seconds to ensure continued accuracy.

Less convincingly perhaps – at least from the watchmaker that had, just a few years earlier in 1969, been acclaimed for its creation of the El Primero mechanical movement – the ad also promises a watch that reflects 'the same workmanship proven in all Zenith television and audio equipment'; the Time Command was launched four years after Zenith the US television/audio brand had acquired a controlling stake in Zenith the Swiss watch brand.

YEAR OF RELEASE 1976

MOVEMENT Zenith 532 quartz module

RELATIVE VALUE ★ ★ ★

NOTABLE FEATURE The small red digital display showed seconds on request

ZODIAC ASTRO

With the influence of the space race on design at its most intense – 1969, of course, being the year of the first moon landing – this automatic watch from Zodiac brings a sense of literal spaciness to its 'mystery' dial. By mounting the hour, minute and second indicators each on their own transparent discs, they are seen to float, rather than, as is more conventional, visibly meet at the centre. The strong use of colour – the pale-blue circle set against an azure surround – serves to add to the otherworldly mood.

The piece somewhat harks back to the founding of the company: Zodiac was established in Le Locle, Switzerland, by Ariste Calame in 1882 (the Zodiac brand was introduced in 1908) with the intention of producing only specialist watches. In 1953 Zodiac introduced the Sea Wolf, one of the first purpose-built diving watches, and in 1968 the Dynatron, believed to be the first Swiss electronic watch.

YEAR OF RELEASE 1969

MOVEMENT Zodiac automatic calibre 88D

RELATIVE VALUE ★ ★ ★

NOTABLE FEATURE Seconds displayed via a red dot 'satellite'

ABOUT THE COLLECTOR

Mitch Greenblatt bought his first vintage watch in 1998. That lit the fuse to what would end a career as an illustrator and ignite a passion for vintage watch collecting, dealing, blogging and, ultimately, designing and producing unique watches for his own watch brand, Xeric.

Mitch's first online presence was as an early eBay dealer selling some of the watches he collected, and quickly became official when, in 1999, *Vogue* contacted him, wanting to write about reselling these rare collectible pieces from the 1960s and 1970s. In order to write about him, they needed a brand to talk about, and that was the birth of Watchismo.com. A Cool Hunting video documented Mitch's collection in 2006, which led to the launch of his 'Watchismo Times' blog and the introduction of new watches at his online store. In 2007, Mitch was approached by the French watch brand Lip, which wanted to break into the US market with reissues of its most popular models from the 1970s.

After years of offering unique timepieces from brands around the world,

ACKNOWLEDGMENTS

the team created their own brand, Xeric watches. They have since launched more than a dozen successful Kickstarter campaigns, including the most funded timepiece in Kickstarter history. Xeric deconstructs the traditions of watch design and pushes the boundaries of how to display time on a mechanical watch.

Mitch now shares his growing collection on Instagram and other social venues as @horolovox.

Watchismo has since acquired and merged into Watches.com, where a long-standing mission of providing unique and affordable modern timepieces is perfectly captured by the motto 'Time to be Different'.

Somewhere out there is the one I'd like to thank most of all: a sexy vintage watch that caught my eye at London's Portobello Market one day in 1998, and that I immediately needed, despite never having collected anything. The watch set me on fire; it had my utmost attention. It was a cocksure block of steel with orange and brown bits, and somehow it also managed wholly to capture the essence of my childhood in 1970s America, both the good and the bad of it. I had to end my trip empty-handed, as I was out of money, but it launched a quest that continues to this day, because I always gravitate to strange and unusual designs in my collecting. In a sense, one elusive watch paved the way for a life of collecting, studying and writing about vintage watches, befriending a world of connoisseurs and scholars, and now finally fulfilling my dream of launching my own watch designs. For all of this, I direct my gratitude into space, hoping it reaches that long-lost mystery watch from Portobello Market, MG.

'For him, it wasn't about price or provenance, but rather about how interesting the watch was. Mitch was then, as he is now, more or less bored by the "traditional" vintage watches out there which are so ubiquitous and often uninspired.'

Ariel Adams, 'A Blog To Watch'

A CHRONOLOGY OF WATCHMAKING

c. 1450 BCE: The first known sundial is employed, in Egypt.

1410 CE: The mainspring is developed, allowing for the first portable clocks.

1492: The first mechanical pocket watches begin to appear.

1541: The first public clock tower is unveiled, at Hampton Court Palace, London.

1554: Thomas Bayard, a Frenchman, opens for business as the first watchmaker in Geneva – the future epicentre of high-end watchmaking.

1581: Galileo discovers the properties of the pendulum.

1610: The first timepiece to be given a protective glass cover is made, better allowing for portability.

1635: Paul Viet introduces the first enamel dials.

1650: The philosopher Blaise Pascal is, perhaps, the first man to wear his watch on his wrist, rather than carry it in a pocket.

1656: Christian Huygens builds on the principles outlined by Galileo to create the first accurate pendulum clocks.

1675: Christian Huygens invents the spiral balance spring. Repeater watches are developed in England.

1705: In England Nicolas Facio, in partnership with watchmakers Peter and Jacob de Beaufré, makes the first watch to use jewels as long-lasting bearings for the movement.

1715: George Graham introduces the deadbeat escapement.

1720: George Graham builds a clock with pendulum and weights capable of indicating the quarter second.

1741: Antoine Thiout outlines the principles underpinning a possible minute-repeater watch, the first of which is believed to have been made by Thomas Mudge.

1757: Thomas Mudge invents the free lever escapement, now used in virtually all mechanical watches.

1766: John Harrison is finally awarded part of the prize money for his H4 marine timekeeper – 52 years after the competition to devise it was launched.

1774: Alexander Cumming invents the gravity escapement.

1775: Jean-Antoine Lepine invents the flat calibre, using bridges, allowing the first compact watches to be constructed.

1776: Jean-Moise Pouzait invests the first watch with independent seconds display.

1784: Thomas Earnshaw improves the accuracy of chronometers by devising a new spring detent escapement.

1790: Pierre Jaquet-Droz and Jean-Frédéric Leschot record the making of possibly the first watch attached to a strap.

1795: Abraham-Louis Breguet invents the tourbillon.

1798: Abraham-Louis Breguet invents the constant force escapement.

1808: In Connecticut, USA, Eli Terry produces 4,000 clocks at $4 each – the first mass-production timepieces.

1822: Nicolas Rieussec invents the seconds chronograph.

1830: Abraham-Louis Breguet introduces the first keyless winding mechanism; other makers follow with their own methods.

1831: Joseph Thaddeus Winnerl unveils the first split-seconds chronograph.

1839: Watch manufacturer Vacheron & Constantin develops a full set of machine tools for the production of its watches, heralding an era of mass production and interchangeable parts.

1840: Alexander Bain invents the first electric clock.

1844: Adolphe Nicole files the first patent for a mechanism that returns a chronograph hand to zero.

1860: The Observatoire de Neuchâtel, founded in 1858, issues the first rating certification for watches.

1867: Georges Frederic Roskopf launches the first watches produced to be 'affordable by all purses'.

1870: Engineer Sandford Fleming proposes the first international time zones.

1875: Calcium sulphate is discovered as the first luminescent material for watch dials and hands.

1880: François Borgel creates the first screwed case, and thus pioneers the water-resistant watch.

1883: The Greenwich Meridian in London is chosen as the prime meridian for the new Universal Time zone system.

1889: The first known patent for a wristwatch is filed in Switzerland.

1892: Auguste Verneuil invents synthetic jewels for use in watch movements.

1907: LeCoultre unveils the first watch movement designed to be super-thin, at just 1.38 mm (5/100 in.) deep.

1909: The first patent for a wrist-worn chronometer is filed.

1913: The basis of most watch cases to come, stainless steel – an alloy of iron, chromium and nickel – is invented by Harry Brearley.

1914: Eterna launches the first mass-production watch with an alarm.

1916: The first patents for a watch capable of recording time to one-fiftieth and one-hundredth of a second are filed by Heuer.

1919: The first compensating balance spring in nickel-steel alloy is made.

1922: In a bid to improve water-tightness, the first wristwatch without crown or stem is introduced by John Harwood.

1924: John Harwood files a patent for the first self-winding wristwatch using an oscillating weight.

1925: Patek Philippe launches the first wristwatch with a perpetual calendar.

1930: The first tourbillon wristwatch is launched.

1931: Louis Cottier invents the first watch showing Universal Time across 29 world cities.

1933: Breitling files a patent for the first two push-button chronograph; Porte-Echappement Universel invents the Incabloc shock absorption system.

1943: Isidor Rabi proposes the use of oscillating atoms to create super-precise clocks.

1949: The first atomic clock is created in the USA.

1952: Elgin and Lip collaborate to create the first electric watch with contacts.

1964: The first electronic watch is launched by Ebauches.

1966: The first quartz wristwatch is prototyped.

1967: The first analogue watch with a quartz movement is unveiled by the Centre Electronique Horloger.

1969: The first self-winding chronographs are launched by Breitling, Zenith, Burer and Heuer; the first quartz watch with LED display is launched too.

1986: Audemars Piguet launches the first series-production self-winding tourbillon wristwatch.

1999: Omega launches the coaxial escapement, invented by George Daniels.

2000: Silicon is used for a wristwatch escapement for the first time.

2005: Seiko launches the first watch with an electromagnetic escapement, dubbed the Spring Drive Kinetic.

ILLUSTRATION CREDITS

Photographs © 2020 Mitch Greenblatt @horolovox horolovox.com except for those listed below.

page 50: Getty 90756517. Picture credit: Science & Society Picture Library / Getty
page 51: Mary Evans 10988297. Picture credit: © Illustrated London News Ltd / Mary Evans
page 99: Alamy B9FEJH. Picture credit: nagelestock.com / Alamy Stock Photo
page 175 below right: Getty 90736844. Picture credit: Science & Society Picture Library /Getty
page 175 above right: Mary Evans 10143024. Picture credit: Mary Evans Picture Library
page 176: Getty 90739422. Picture credit: Science & Society Picture Library / Getty

pages 53 and 98–101: Illustrations from *Mid-Century Modern Complete*. Photos courtesy of Wright, Chicago IL, http://wright20.com

Josh Sims is a UK-based journalist and editor, contributing to many publications at home and abroad. He is also the author of several books on matters of style.

Mitch Greenblatt has collected and sold vintage watches since 1998. He is the co-founder and CEO of Watches.com and has designed and produced unique watches for his own brand Xeric since 2013.

Tyler Little has been photographing people, places and products for the past 19 years. He currently works out of the San Francisco Bay Area.

On the cover, front, left: Hamilton; front, right: Enicar Sherpa 320; back: Desotos 'Cuffbuster' Chronograph

Retro Watches © 2020 Thames & Hudson Ltd, London

Text © 2020 Josh Sims

Photographs © 2020 Mitch Greenblatt @horolovox horolovox.com unless otherwise stated

All Rights Reserved. No part of this publication may be reproduced or transmitted in any form or by any means, electronic or mechanical, including photocopy, recording or any other information storage and retrieval system, without prior permission in writing from the publisher.

First published in 2020 in the United States of America by Thames & Hudson Inc., 500 Fifth Avenue, New York, New York 10110

www.thamesandhudsonusa.com

Library of Congress Control Number 2019940754

ISBN 978-0-500-02296-2

Printed and bound in China by C&C Offset Printing Co. Ltd